T0254692

Practical Hydraulics and Water Resources Engineering

THIRD EDITION

Practical Hydraulics and Water Resources Engineering

THIRD EDITION

Melvyn Kay

CRC Press
Taylor & Francis Group
Boca Raton London New York

CRC Press is an imprint of the
Taylor & Francis Group, an **informa** business

CRC Press
Taylor & Francis Group
6000 Broken Sound Parkway NW, Suite 300
Boca Raton, FL 33487-2742

© 2017 by Taylor & Francis Group, LLC
CRC Press is an imprint of Taylor & Francis Group, an Informa business

No claim to original U.S. Government works

Printed on acid-free paper
Version Date: 20161021

International Standard Book Number-13: 978-1-4987-6195-6 (Paperback)

Library of Congress Cataloging-in-Publication Data

Names: Kay, Melvyn. | Kay, Melvyn. Practical hydraulics.
Title: Practical hydraulics and water resources engineering / Melvyn Kay.
Other titles: Practical hydraulics
Description: Third edition. | New York, NY : Routledge, 2017. | Previous
edition: Practical hydraulics / Melvyn Kay (London ; New York : Taylor &
Francis, 2008). | Includes bibliographical references and index.
Identifiers: LCCN 2016027431| ISBN 9781498761956 (pbk.) | ISBN 9781498761963
(e-book)
Subjects: LCSH: Hydraulics. | Hydraulic engineering.
Classification: LCC TC160 .K38 2017 | DDC 620.1/06--dc23
LC record available at https://lccn.loc.gov/2016027431

Visit the Taylor & Francis Web site at
http://www.taylorandfrancis.com

and the CRC Press Web site at
http://www.crcpress.com

Contents

4 Pipes   107

9   Water resources planning and management   317

Preface

Who wants to know about hydraulics? When I asked that question as I was finalising the second edition of this book in 2007, it was my 6-year-old daughter who wanted to know why water swirls as it goes down the plug hole when she has a bath and why it always seems to go in the same direction. She is now 16 and has developed other interests! But many people from a wide range of backgrounds are now engaging in careers linked to water and water use: in public water supply, agriculture, irrigation, energy, environment and sustainable development. Those of you involved in some aspect of water management will need a good understanding of basic hydraulics to do your job. So this book is aimed at you. There is no need for you to become a hydraulic engineer but you are most likely to work with engineers, and so you will need to ask the right questions, and know when you are being given sensible answers. You will need to understand how engineers set about the vital role of harnessing and controlling water. This book sets out to explain how water flows in pipes, channels and pumps, and how it is controlled and managed.

Most traditional hydraulics books are written by engineers for engineers and so they tend to assume a strong background in maths and physics and are rather off putting for those who are not from an engineering background. This book is still written by an engineer, but for many years, I taught hydraulics to young professionals from a wide range of backgrounds and from this I have developed what I think is a more 'novel' approach to understanding hydraulics, which is simple and practical. I have tried to avoid (most of) the maths and physics that many of you may not be so keen on. Instead, I use lots of 'stories' that I have gathered over many years to help you quickly grasp important water engineering principles and practices.

It would have been easier to write a 'simple', descriptive book on hydraulics by omitting the more complex ideas of water flow but this would have been simplicity at the expense of reality. It would be like writing a cookbook with recipes rather than examining why certain things happen when ingredients are mixed together. So I have tried to cater for various tastes. At one level, this book is descriptive and provides a qualitative understanding of hydraulics. At another level, it is more rigorous and quantitative. These are the more mathematical bits for those who wish to go that extra step. It was the physicist, Lord

Kelvin (1824–1907) who said that it is essential to put numbers on things if we are really going to understand them. So if you are curious about solving problems, I have included a number of worked examples, as well as some of the more interesting formula derivations and put them into boxes in the text so that you can spot them easily, and avoid them if you wish.

This is the third edition of *Practical Hydraulics* and rather than go deeper into hydraulics I have, in addition to taking out the errors in early editions that readers have kindly pointed out to me, added two new chapters which show how engineers apply the principles in practice in the world of water.

Chapter 8 is about water resources engineering; how engineers work and think, and how they apply their skills to water supply, wastewater treatment, irrigation and to engineering our rivers and the aquatic environment.

Chapter 9 is about water resources planning and management. Since the second edition, water has moved to the centre of world attention as never before. This has been driven by the extremes of floods and droughts that we are now experiencing and of course the potential impacts of climate change. In 2015, the World Economic Forum identified water as the number one global risk facing humanity. And in 2016, the United Nations recognised that water is at the heart of all aspects of development and sustainable growth and so deserves special attention by all member nations. So a chapter on water resources planning and management seems most apt. But this is becoming a highly complex issue and many textbooks are now addressing this. I have chosen to try and introduce the reader to this important subject and highlight some of the current thinking about water resources and the direction in which we are going. For example, do we have enough water to sustain our way of life in the future? I have also included a most important section at the end of this chapter which I call Water Myths. Many people are now dabbling in water and have limited understanding of how it behaves in practice. This can lead to poor decision-making. Is water really finite? What does water-use efficiency actually mean? And is wastewater really a 'new' water resource? All is revealed in Chapter 9.

One final point to bear in mind. Developing a qualitative understanding of hydraulics and solving problems mathematically are two different skills. Many people achieve a good understanding of water behaviour but then get frustrated because they cannot easily apply the maths. This is a common problem and it requires a skill that can only be acquired through practice – hence the reason for the worked examples in the text. I have also included a list of problems at the end of each chapter for you to try out your new skills. It does help to have *some* mathematical skills – basic algebra should be enough to get you started.

So enjoy learning about hydraulics and I hope my book enriches your career in the world of water.

Melvyn Kay
October 2016

Acknowledgements

I would like to make special mention of two books which have greatly influenced my writing of this text. The first is *Water in the Service of Man* by HR Vallentine, published by Pelican Books Ltd in 1967. The second is *Fluid Mechanics for Civil Engineers* by NB Webber first published in 1965 by E&FN Spon Ltd. These may seem old references but they are excellent texts and the basic principles of physics do not really change. Unfortunately, both are now out of print but copies can still be found via Amazon.

I would like to acknowledge my use of the method described in *Handbook of Hydraulics for the Solution of Hydrostatic and Fluid Flow Problems* by HW King and EF Brater published in 1963 for the design of channels using Manning's equation in Chapter 5.

I am also grateful for ideas I obtained from *The Economist* on the use of boundary drag on swim suits (Section 3.10) and from *New Scientist* on momentum transfer and the hydrodynamics of cricket balls (Sections 1.10 and 3.12).

I would like to thank the following people and organisations for permission to use photographs and diagrams:

Chadwick A, Morfett J, and Borthwick M. 2004. *Hydraulics in Civil and Environmental Engineering.* 4th edition. E & FN Spon, London for Figure 5.24b.

David Thomas, Chief Engineer, Middle Level Commissioners, Norfolk, UK for Figure 8.1.

FC Concrete Ltd, Derby UK for Figure 6.14.

Fox J. 1977. *An Introduction to Engineering Fluid Mechanics.* The MacMillan Press Ltd, London, UK for Figure 5.28.

Fraenkel PL. 1986. *Water Lifting Devices.* Irrigation and Drainage Paper No 43 Food and Agriculture Organisation, Rome for Figures 7.5, 7.11, and 7.21a and b.

Hess T. Cranfield University for figure and information provided for Figure 9.7.

Hydraulics Research Wallingford. 1983. *Charts for the Hydraulic Design of Channels and Pipes*. 5th edition. Thomas Telford Ltd, UK for Figure 4.8.

IPTRID-FAO. 2000. Treadle pumps in Africa. Knowledge Synthesis paper No1 for Figure 7.3.

ITT Lowara Pumps Ltd for Figure 7.19.

McGregor D. Royal Holloway, University of London for Figure 9.1

Open University Oceanography COURIS Team. 1995. *Waves, Tides and Shallow Water Processes*. Butterworth and Heineman 1995 for Figure 10.1.

Photographer Tom Brabben for Figure 7.3b

US Navy photo by Ensign John Gay for Figure 5.15c.

Vallentine HR. 1967. *Water in the Service of Man*. Penguin Books Ltd, Harmondsworth, UK for Figures 2.7 and 7.2 a, b and c.

Webber NB. 1971. *Fluid Mechanics for Civil Engineers*. E & FN Spon Ltd, London, UK for Figures 7.10b and 7.21c.

Wijngaarden MA. For picture of modern Archimedes' screw in a pumping station in Kinderdijk, The Netherlands for Figure 8.3.

Figures 2.18, 2.19, 8.2, 8.4, 8.7, 8.8, 8.9 and 8.10 are reproduced with permission of Shutterstock©

Author

Melvyn Kay is a chartered civil engineer with over 40 years experience in hydraulics and water resources management, particularly in irrigation for food production. He spent 10 years with consulting engineers in the Middle East and Africa on major hydraulic works before joining Cranfield University where he became a senior lecturer teaching hydraulics and irrigation engineering to international postgraduate students, many of them were from 'non-engineering' backgrounds. Then, he served as a director of Training and Consultancy Services, and was responsible for the development of international training and consultancy contract work at Cranfield University. He now works as an independent consultant on water for food production, capacity development, knowledge transfer in agricultural water development and management for various agencies, such as FAO (Food and Agriculture Organization), GWP (Global Water Partnership) and the World Bank.

Chapter 1

Some basic mechanics

Water is now at the centre of world attention as never before and more professionals from all walks of life are engaging in careers linked to water – in public water supply and waste treatment, agriculture, irrigation, energy, environment, amenity management and sustainable development. This book offers an appropriate depth of understanding of basic hydraulics and water resources engineering for those who work with civil engineers and others in the complex world of water resources development, management and water security. It is simple, practical and avoids (most of) the maths in traditional textbooks. Lots of excellent 'stories' help readers to quickly grasp important water principles and practices.

This third edition is broader in scope and includes new chapters on water resources engineering and water security. Civil engineers may also find it a useful introduction to complement the more rigorous hydraulics textbooks.

1.1 INTRODUCTION

This is a reference chapter rather than one for general reading, but it will be useful as a reminder about the physical properties of water and for those who want to revisit some basic physics which is directly relevant to the behaviour of water.

1.2 DIMENSIONS AND UNITS

To understand hydraulics, it is essential to put numerical values on such things as pressure, velocity and discharge in order for them to have meaning. It is not enough to say the pressure is high or the discharge is large; some specific value needs to be given to quantify it. Also, providing just a number is quite meaningless – it needs a unit. To say a pipeline is 6 long is not enough. It might be 6 cm, 6 m or 6 km. So dimensions must have numbers and numbers must have units to give them some meaning.

Different units of measurement are used in different parts of the world. The United States still uses the foot, pound and second system (known as the FPS system), and to some extent, this system still exists in the United Kingdom. In continental Europe, two systems are in use depending on which branch of science you are in. There is the centimetre–gramme–second (CGS) and also the metre–kilogramme–second (MKS) system. All very confusing and so in 1960, after years of discussion and finally international agreement, a new International System of Units (known as SI system based on the French: Systeme International d'Unites) was established. Not everyone has switched to this, but most engineers now accept the system and so in order to avoid any further confusion, we shall use this throughout this book.

SI system is not a difficult system to grasp and it has many advantages over other systems. It is based mainly on the MKS system. All length measurements are in metres, mass is in kilogrammes and time is in seconds (Table 1.1). SI units are simple to use and their big advantage is they can help to avoid much of the confusion which surrounds the use of other units. For example, it is quite easy to confuse mass and weight in both FPS and MKS systems. In FPS, both are measured in pounds, and in MKS, they are measured in kilogrammes. Any mix-up between them can have serious consequences for the design of engineering works. In the SI system, the difference is clear because they have different units – mass is measured in kilogrammes, whereas weight is measured in Newtons. More about this is explained later in Section 1.7.

Note that there is no mention of centimetres in Table 1.1. Centimetres are part of the CGS system and play no part in hydraulics or in this text. Millimetres are acceptable for very small measurements and kilometres for long lengths – but *not* centimetres. The litre (L) is also not officially an SI unit, though most water supply engineers talk about megalitres when referring to water volumes. So even the SI system has its idiosyncrasies.

Every measurement must have a unit so that it has meaning. The units chosen do not affect the quantities measured and so, for example, 1.0 m is exactly the same as 3.28 feet. However, when solving problems, all the measurements used must be in the same system of units. If we mix them up in a formula (e.g. centimetres or inches instead of metres, or minutes instead of seconds), the answer will be meaningless. Some useful units which are derived from the basic SI units are included in Table 1.2.

Table 1.1 Basic SI units of measurement

Measurement	Unit	Symbol
Length	Metre	m
Mass	Kilogramme	kg
Time	Second	s

Table 1.2 Some useful derived units in this SI system

Measurement	Unit	Measurement	Unit
Area	m^2	Force	N
Volume	m^3	Mass density	kg/m^3
Velocity	m/s	Weight density	N/m^3
Acceleration	m/s^2	Pressure	N/m^2
Viscosity	kg/ms	Momentum	kgm/s
Kinematic viscosity	m^2/s	Energy	Nm/N

1.3 VELOCITY AND ACCELERATION

Most people use the terms *velocity* and *speed* to mean the same. But scientifically they are different. Speed is the rate at which some object is travelling and is measured in metres/second (m/s), but this does not tell you in which direction the object is going. Velocity is speed plus direction. It defines movement in a particular direction and is also measured in metres/second (m/s). In hydraulics, it is useful to know in which direction water is moving and so the term velocity is used instead of speed. When an object travels a known distance in a given time, we can calculate the velocity as follows (see example in Box 1.1):

$$\text{velocity} = \frac{\text{distance (m)}}{\text{time (s)}}.$$

Acceleration describes change in velocity. When an object's velocity is increasing, it is *accelerating*; when it is slowing down, it is *decelerating*.

BOX 1.1   EXAMPLE: CALCULATING VELOCITY AND ACCELERATION

An object is moving along at a steady velocity and it takes 150 s to travel 100 m. Calculate the velocity.

$$\text{velocity} = \frac{\text{distance (m)}}{\text{time (s)}} = \frac{100}{150} = 0.67 \text{ m/s}.$$

Calculate the acceleration when an object starts from rest and reaches a velocity of 1.5 m/s in 50 s.

$$\text{acceleration} = \frac{\text{change in velocity (m/s)}}{\text{time (s)}} = \frac{1.5 - 0}{50} = 0.03 \text{ m/s}^2.$$

Acceleration is measured in metres/second/second (m/s²). If the initial and final velocities are known as well as the time taken for the velocity to change, we can calculate the acceleration as follows:

$$\text{acceleration (m/s}^2) = \frac{\text{change in velocity (m/s)}}{\text{time (s)}}.$$

1.4 FORCES

Force is not a word that we can easily describe in the same way as we can describe some material object. Instead, we talk about a pushing or pulling action and so we say what a force will do and not what it is. Using this idea, if we apply a force to some stationary object it will begin to move (Figure 1.1). If we apply it for long enough, then the object will begin to move faster, that is, it will accelerate. The same applies to water and to other fluids as well. It may be difficult to think of pushing water, but if it is to flow along a pipeline or a channel, a force will be needed to move it. So one way of describing force is to say that *a force causes movement.*

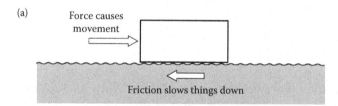

Figure 1.1 Forces and friction. (a) Friction resists movement. (b) Trying to 'swim in a frictionless fluid'.

1.5 FRICTION

Friction is the name we give to the force which resists movement and so causes objects to slow down (Figure 1.1a). It is an important aspect of all our daily lives. Without friction between our feet and the ground surface, it would be impossible to walk, and we are reminded of this each time we step onto ice or some smooth oily surface. We would not be able to swim if water was frictionless. Our arms would just slide through the water and we would not make any headway – just like children trying to 'swim in a sea of plastic balls' in the playground (Figure 1.1b).

But friction can also hinder our lives. In car engines, friction between the moving parts would quickly create heat and the engine would seize up. Oil lubricates the surfaces and reduces the friction.

Friction also occurs between flowing water and the internal surface of a pipe or the bed and sides of a channel. Indeed, much of pipe and channel hydraulics is about predicting this friction force so that we can select the size of pipe or channel to carry a given discharge (Chapters 4 and 5).

Friction is not only confined to boundaries, there is also friction inside fluids (internal friction) which makes some fluids flow more easily than others. The term *viscosity* is used to describe this internal friction (see Section 1.11.3).

1.6 NEWTON'S LAWS OF MOTION

Sir Isaac Newton (1642–1728) was one of the early scientists who studied forces and how they cause movement. His work is now enshrined in three basic rules, known as *Newton's Laws of Motion*. They are very simple laws, and at first sight, they appear so obvious, they seem hardly worth writing down. But they form the basis of all our understanding of hydraulics (and movement of solid objects as well) and it took the genius of Newton to recognise their importance.

1.6.1 Law 1: Forces cause objects to move

First, imagine what happens with solid objects. A block of wood placed on a table will stay there unless you push it (i.e. a force is applied to it). Equally, if it is moving, it will continue to move unless some force (e.g. friction) causes it to slow down or to change direction. So forces are needed to make objects move or to stop them. This same law applies to water.

1.6.2 Law 2: Forces cause objects to accelerate

This law builds on the first and describes how forces cause objects to accelerate (Figure 1.2a). The size of the force needed depends on the size of the

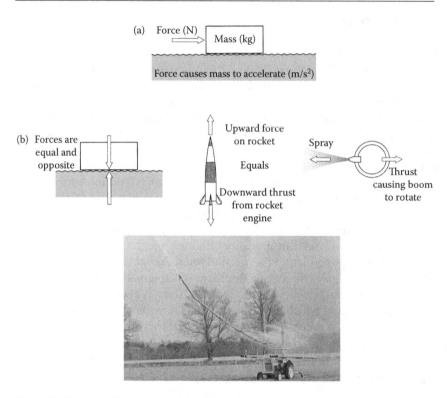

Figure 1.2 Newton's laws of motion. (a) Newton's 2nd law. (b) Examples of Newton's 3rd law.

block (its mass) and how quickly you want it to accelerate. The larger the block and the faster it must go, the larger must be the force. Water behaves in the same way. Engineers want to know about the force needed to push water along a pipeline or channel. Newton established the equation for calculating the force, linking it to mass and acceleration (see example in Box 1.2):

force (N) = mass (kg) × acceleration (m/s^2).

The unit of force is derived from multiplying mass and acceleration, that is, kg m/s^2. But this is rather a cumbersome unit and the SI system calls this a *Newton* (N) in recognition of Sir Isaac Newton's contribution to our understanding of mechanics. A force of 1 N is defined as the force needed to cause a mass of 1 kg to accelerate at 1 m/s^2. This is not a large force. An apple held in the palm of your hand weighs approximately 1 N – an interesting point as it was supposed to have been an apple falling on Newton's head that set off his thoughts on forces, gravity and motion.

BOX 1.2   EXAMPLE: CALCULATING FORCE
USING NEWTON'S SECOND LAW

A mass of 3 kg is to be moved from rest to reach a velocity of 6 m/s in 10 s.
Calculate the force needed.

First calculate acceleration:

$$\text{acceleration (m/s}^2) = \frac{\text{change in velocity (m/s)}}{\text{time (s)}}$$

$$\text{acceleration} = \frac{6}{10} = 0.6 \text{ m/s}^2.$$

Use Newton's second law:

force = mass × acceleration

= 3 × 0.6 = 1.8 N.

So a force of 1.8 N is needed to move a mass of 3 kg to a velocity of 6 m/s in 10 s.

Working in Newtons in engineering and hydraulics will produce some very large numbers and so to overcome this, forces are usually measured in kilo-Newtons (kN):

1kN = 1,000 N.

1.6.3 Law 3: For every force there is always an equal and opposite force

To understand this simple but vitally important law, again think of the block of wood sitting on a table (Figure 1.2b). The block exerts a force (its weight) downwards on the table, but the table also exerts an equal but opposite upward force on the block. If it did not, the block would drop down from the table under the influence of gravity. So there are two forces, exactly equal in magnitude but in opposite directions and so the block does not move.

The same idea can be applied to moving objects as well. In earlier times, it was thought that objects, such as arrows, were propelled forward by the air rushing in behind them. This idea was put forward by the Greeks, but it failed to show how an object could be propelled in a vacuum as is the case when a rocket travels into space. What in fact happens is the downward thrust of the burning fuel creates an equal and opposite thrust which pushes the rocket upwards (Figure 1.2c). Newton helped to discredit the Greek idea by setting up an experiment which showed that, rather than encouraging an arrow to fly faster, the air flow actually slowed it down because of the friction between the arrow and the air.

Another example of Newton's third law occurs in irrigation where rotating booms spray water over crops (Figure 1.2d). The booms are not powered by a motor, rather they are driven by the reactive force from water jets. As water is forced out of the nozzles along the boom, it creates an equal and opposite force on the boom which causes it to rotate. The same principle is used to drive the rotating pipes which spray water over coke beds in trickle filters in wastewater treatment plants (see Section 8.4.4).

1.7 MASS AND WEIGHT

There is often confusion between mass and weight and this has not been helped by past systems of units and also the way we use the terms in everyday language. Mass and weight have very specific scientific meanings and for any study of water it is essential to have a clear understanding of the difference between them.

Mass refers to an amount of matter or material. It is a constant value and is measured in kilogrammes (kg). A specific quantity of matter is often referred to as an *object*. Hence, the use of this term in the earlier description of Newton's laws.

Weight is a force. It is a measure of the force of gravity on an object. On the Earth, there are slight variations in gravity from place to place. These variations can be very important for athletes. They can affect how fast or how high they can jump and make that slight difference which means a world record. But in engineering terms and for most of us, minor changes have no effect on our daily lives.

Gravity on the Moon is much less than it is on the Earth. So if we take an object to the Moon, its mass is still the same, but its weight is much less. As weight is a force, it is measured in Newtons and this clearly distinguishes it from mass which is measured in kilogrammes.

Newton's second law enables us to calculate the weight of a given mass. In this case, the acceleration term is the acceleration resulting from gravity. This is the acceleration that any object experiences when it is dropped and allowed to fall to the Earth's surface. Objects dropped do experience different rates of acceleration but this is because of the air resistance – hence the reason why a feather falls more slowly than a stone. But if both were dropped at the same time in a vacuum, they would fall (accelerate) at the same rate.

For engineering purposes, acceleration due to gravity is assumed to have a constant value of 9.81 m/s^2. This is usually referred to as the *gravity constant* and denoted by the letter *g*. Newton's second law provides the link between mass and weight as follows (see example in Box 1.3):

weight (N) = mass (kg) × gravity constant (m/s^2).

**BOX 1.3 EXAMPLE: CALCULATING
THE WEIGHT OF AN OBJECT**

Calculate the weight of an object whose mass is 5 kg.

Use Newton's second law:

weight = mass × gravity constant
= 5 × 9.81 = 49.05 N.

Sometimes engineers assume the gravity constant is 10 m/s^2 because it is easier to multiply by 10 and the error involved in this is not significant in engineering terms.

In this case:

weight = 5 × 10 = 50 N.

Confusion between mass and weight occurs because of the way we use the words in our everyday lives. When visiting a shop, we might ask for 5 kg of potatoes, which are duly weighed out on a weigh balance graduated in kilogrammes. Costs are quoted in £ per kilogramme. But to be strictly correct, we should ask for 50 N of potatoes as the balance is measuring the *weight* of the potatoes (i.e. the force of gravity) and not their mass. But, because gravity is constant all over the world (or nearly so for engineering purposes), the conversion factor between mass and weight is a constant value of 9.81. So we tend to stick with asking for kilogrammes rather than Newtons. If shopkeepers were to change their balances to read in Newtons to resolve a scientific confusion, engineers and scientists might be happy but no doubt a lot of shoppers would not be so happy!

1.8 SCALAR AND VECTOR QUANTITIES

Measurements in hydraulics are either called *scalar* or *vector* quantities. Scalar measurements only indicate magnitude. Examples of this are mass, volume, area and length. So if there are 120 boxes in a room and they each have a volume of 2 m^3, both the number of boxes and the volume of each are scalar quantities. Scalar quantities can be added together using the rules of arithmetic. Thus, 5 boxes and 4 boxes can be added to make 9 boxes and 3 m and 7 m can be added to make 10 m.

Vectors have direction as well as magnitude. Examples of vectors include force and velocity. It is just as important to know which direction forces are pushing and water is moving as well as magnitude.

1.8.1 Dealing with vectors

Vectors can also be added together providing their direction is taken into account. Adding (or subtracting) two or more vectors results in another single vector called the *resultant* and the vectors that make up the resultant are called the *components*. If two forces, 25 and 15 N, are pushing in the same direction then their resultant is found simply by adding the two together, that is, 40 N (Figure 1.3a). If they are pushing in opposite directions, then their resultant is found by subtracting them, that is, 10 N. So one direction is considered positive and the opposite direction negative for the purposes of combining vectors.

But forces can also be at an angle to each other and in such cases a different way of adding or subtracting them is needed – a *vector diagram* is used for this purpose. This is a diagram, drawn to a chosen scale to show both the magnitude and the direction of the vectors and hence the magnitude of the resultant vector. An example of how this is done is shown (Box 1.4).

Vectors can also be added and subtracted mathematically, but knowledge of trigonometry is needed. For those interested in this approach, it is described in most basic books on maths and mechanics.

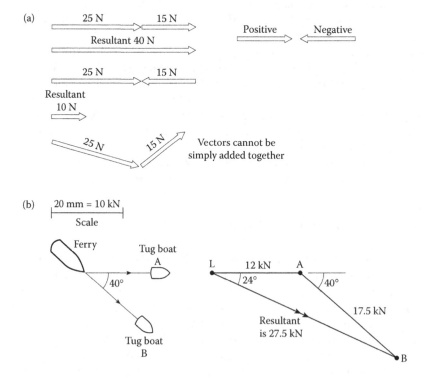

Figure 1.3 Adding and subtracting vectors. (a) Calculating the resultant. (b) The tug boat problem (Box 1.4).

BOX 1.4   EXAMPLE: CALCULATING THE RESULTANT FORCE USING A VECTOR DIAGRAM

Two tug boats, A and B, are pulling a large ferry boat into a harbour. Tug A is pulling with a force of 12 kN, tug B with a force of 17.5 kN, and the angle between the two ropes is 40° (Figure 1.3b). Calculate the resultant force and show the direction in which the ferry boat will move.

First draw a diagram of the ferry and the two tugs. Then, assuming a scale of 20 mm equals 10 kN (this is chosen so that the diagram fits conveniently onto a sheet of paper) draw the 12 kN force to scale, that is, the line LA. Next, draw the second force, 17.5 kN, to the same scale but starting the line at A and drawing it at an angle of 40° to the first line. This 'adds' the second force to the first one. The resultant force is found by joining the points L and B. The length of the line LB in mm can be converted into kN using the chosen scale. The ferry will be pulled in the direction of the resultant force. So the resultant force is 27.5 kN and the ferry will move in a direction 24° to the line of Tug A.

The triangle drawn in Figure 1.3b is the *vector diagram* and shows how two forces can be added together. There are only three forces in this problem which is called a *triangle of forces*. However, it is possible to add together many forces using the same method. In such cases, the diagram is referred to as a *polygon of forces*.

1.9 WORK, ENERGY AND POWER

Work, energy and power are all words used in everyday language, but in engineering and hydraulics, they have very specific meanings.

1.9.1 Work

Work refers to almost any kind of physical activity, but in engineering, work done is about moving objects. A crane does work when it lifts a load against the force of gravity and a train does work when it pulls carriages along a track. But if you hold a large weight for a long period of time, you will undoubtedly get tired and feel that you have done lot of work, but in engineering terms, you have not done any work because nothing moved.

Work done on an object can be calculated as follows:

work done (Nm) = force (N) × distance moved by an object (m).

Work done is the product of force (N) and distance (m) so it is measured in Newton-metres (Nm).

1.9.2 Energy

Energy enables useful work to be done. People and animals require energy to do work. They get this by eating food and converting it into useful energy for work through the muscles of the body. Energy is also needed to make water flow and this is why reservoirs are built in mountainous areas so that the natural energy of water can be used to make it flow downhill to a town or to a hydroelectric power station. Where 'natural' energy is not available, water is pumped to provide the right pressures and flow. In effect, a pump is just a means of adding energy into a water system. This can come from an electric motor or from a diesel or petrol engine. Solar and wind energy are alternatives and so is energy provided by human hands or animals.

The amount of energy needed to do a job is determined by the amount of work to be done. So that:

energy required (Nm) = force (N) × distance (m).

Energy, like work, is measured in Newton-metres (Nm), but the more conventional measurement of energy is *Watt-seconds* (Ws), named after the Scottish engineer James Watt (1736–1819) where:

1 Ws = 1 Nm.

But this is a very small quantity for engineering work and so the preferred unit is *Watt-hours* (Wh) or *kilowatt-hours* (kWh). To change Ws to Wh, multiply both sides of the above equation by 3,600 to change seconds to hours:

1 Wh = 3,600 Nm.

Now multiply both sides by 1,000 to change Wh to kWh:

1 kWh = 3,600,000 Nm
      = 3,600 kNm.

While engineers like to measure energy in kilowatt-hours (kWh), scientists tend to measure energy in Joules (J). This is recognition of the contribution made by the English physicist, James Joule (1818–1889) to our understanding of energy, in particular the conversion of mechanical energy to heat energy (see Section 1.9.3).

So for the record:

1 J = 1 Nm.

However, to avoid using too many units, we will stay with kilowatt-hours in this text. Some everyday examples of energy use include:

- A farmer working in the field uses 0.2–0.3 kWh every day
- An electric desk fan uses 0.3 kWh every hour
- An air conditioner uses 1 kWh every hour

Note that it is important to specify the time period (e.g. every hour, every day) over which the energy is used. Energy used for pumping water is discussed more fully in Chapter 7.

1.9.3 Changing energy

One of the useful things about energy is that it can be changed from one form of energy to another. People and animals are able to convert food into useful energy to drive their muscles. The farmer using 0.2 kWh every day must eat enough food each day to supply this energy need otherwise he would not be able to work properly.

In a diesel pumping set, the form of energy changes several times before it gets to the water. Chemical energy contained within the fuel (e.g. diesel oil) is burnt in a diesel engine to produce mechanical energy. This is converted to useful water energy via the drive shaft and pump (Figure 1.4). So a pumping unit is both an energy converter as well as a devise for adding energy into a water system.

Changing energy from one form to another is not always very efficient. Friction between the moving parts absorbs energy and this is converted into heat energy and is usually lost into the atmosphere. Such losses can be high and also costly as they waste fuel. One of the criteria used to match pumps and power units is to ensure that as much energy as possible goes into the water and maximises the efficiency of energy use (see Chapter 7).

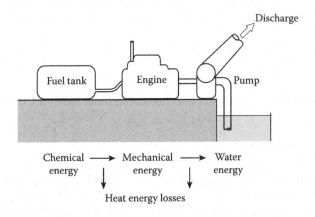

Figure 1.4 Changing energy from one form to another.

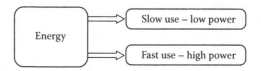

Figure 1.5 Power is the rate of energy use.

1.9.4 Power

Power is often confused with the term energy. They are related but they have different meanings. While energy is the capacity to do useful work, power is the rate at which the energy is used (Figure 1.5).

And so:

$$\text{power (kW)} = \frac{\text{energy (kWh)}}{\text{time (h)}}.$$

Examples of power requirements: a typical room air conditioner has power rating of 1 kW. This means that it consumes 1 kWh of energy every hour it is working. A small electric radiator has a rating of 1–2 kW, and the average person walking up and down stairs has a power requirement of about 70 W.

Energy requirements are sometimes calculated from knowing the equipment power rating and the time over which it is used rather than trying to calculate it from the work done. In this case:

energy (kWh) = power (kW) × time (h).

Horsepower (HP) is still a very commonly used measure of power, but it is not used in this book, as it is not an SI unit. However, for comparison purposes:

1 kW = 1.36 HP.

Power used for pumping water is discussed more fully in Chapter 7.

1.10 MOMENTUM

Applying a force to a mass causes it to accelerate (Newton's second law) and the effect of this is to cause a change in velocity. This means there is a link between mass and velocity and this is called *momentum*. Momentum is another scientific term that is used in everyday language to describe something that is moving – we say that some object or a football game has momentum if it is moving along and making good progress. In engineering terms, it has a specific meaning and it can be calculated by multiplying the mass and the velocity together:

momentum (kgm/s) = mass (kg) × velocity (m/s).

Note that the unit of momentum are a combination of those of velocity and mass.

The following example demonstrates the links between force, mass and velocity. Two blocks, 2 and 10 kg mass, are being pushed along by an identical force of 15 N for 4 s (Figure 1.6a). To keep the example simple, assume there is no friction.

To calculate the momentum of the 2 kg block, first calculate the acceleration using Newton's second law and then calculate the final velocity after the force is applied for 4 s.

force = mass × acceleration
$$15 = 2 \times f$$
$$f = 7.5 \text{ m/s}^2.$$

For every second the force is applied, the block will move faster by 7.5 m/s. After 4 s, it will reach a velocity of

$$4 \times 7.5 = 30 \text{ m/s}.$$

Calculate the momentum of the 2 kg block:

momentum = mass × velocity
$$= 2 \times 30 = 60 \text{ kg m/s}.$$

Now calculate the momentum for the 10 kg block using the same approach:

force = mass × acceleration
$$15 = 10 \times f.$$

And so

$$f = 1.5 \text{ m/s}^2.$$

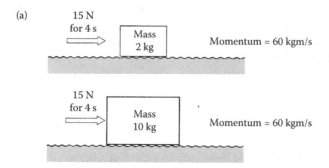

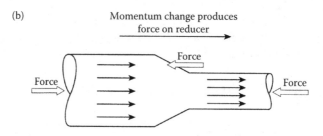

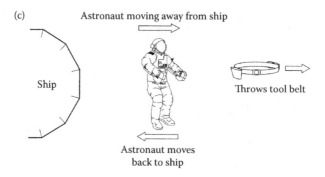

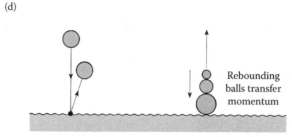

Figure 1.6 Understanding momentum. (a) Momentum for solid objects. (b) Momentum change produces forces. (c) The astronaut's problem. (d) Rebounding balls.

For every second the force is applied, the block will move faster by 1.5 m/s. After 4 s, it will reach a velocity of

$4 \times 1.5 = 6$ m/s.

Now calculate momentum of the 10 kg block:

momentum = mass × velocity
momentum = $10 \times 6 = 60$ kg m/s.

The result is that the same force applied over the same time period to very different blocks produces the same momentum.

Now multiply the force by the time:

force × time = $15 \times 4 = 60$ Ns.

But the unit for Newtons can also be written as kgm/s^2 and so:

force × time = 60 kgm/s.

This is equal to the momentum and has the same units.

Multiplying force by time is called the *impulse* and this is equal to the momentum it creates.

And so we can now write:

impulse = momentum.

And

force × time = mass × velocity.

This is more commonly written as

impulse = change of momentum.

Writing *change in momentum* is more appropriate because an object need not be starting from rest, it may already be moving when it is pushed. In such cases, the object will have some momentum and an impulse would increase (change) it. Also a momentum change need not be just a change in velocity, it can also be a change in mass. If a lorry loses some of its load when travelling at speed, its mass will change. If momentum is maintained, then the lorry will speed up as a result of the smaller mass.

The equation for momentum change becomes

$$\text{force} \times \text{time} = \text{mass} \times \text{change in velocity.}$$

This equation works well for solid blocks, but more steps are needed to apply it to flowing water. Here, we want to look at the rate at which the mass of water is flowing rather than thinking of the flow as a series of discrete solid blocks of water. To do this, we first divide both sides of the equation by time, thus:

$$\text{force} = \frac{\text{mass}}{\text{time}} \times \text{change in velocity.}$$

Mass divided by time is the mass flow in kg/s and so the equation becomes:

$$\text{force (N)} = \text{mass flow (kg/s)} \times \text{change in velocity (m/s).}$$

So applying a force to water, such as with a pump or a turbine, we can create a change in momentum which may increase the mass flow (essentially the discharge) or a change in the water velocity. Equally if we change the momentum by changing the velocity or the flow, such as water flowing around a pipe bend or through a reducer, it creates a force as a direct result of those changes (Figure 1.6b). Engineers will need to make sure these forces are well contained within the pipe system.

Momentum is about forces and velocities, which are vectors, and so the direction in which momentum changes is also important. In the above example (Figure 1.6b), the force is pushing from left to right as the flow is from left to right. This is assumed to be the positive direction. Any force or movement from right to left would be considered negative. So if several forces are involved, they can be added or subtracted to find a single resultant force. Another important point to note is that Newton's third law also applies to momentum. Figure 1.6b shows the force of the reducer on the water and so the force is in the negative direction. Equally it could be drawn in the opposite direction, that is, the positive direction from right to left when it would be the force of the water on the reducer. The two forces are, of course, equal and opposite.

The application of this idea to water flow is developed further in Section 4.11.

Those not so familiar with Newton's laws might find momentum more difficult to deal with than other aspects of hydraulics. To help understand the concept, here are two interesting examples of momentum change which might help.

1.10.1 The astronaut's problem

An astronaut has just completed a repair job on his spaceship and secures his tools on his belt. He then pushes off from the ship to drift in space only to find that his lifeline has come undone and he is drifting further and further away from his ship (Figure 1.6c). How can he get back? One solution is for him to take off his tool belt and throw it as hard as he can in the direction he is travelling – away from the ship. The reaction from this will be to propel him in the opposite direction back to his spaceship. The momentum created by throwing the tool belt in one direction (i.e. mass of tool belt multiplied by velocity of tool belt) will be matched by a momentum change in the opposite direction (i.e. mass of spaceman multiplied by velocity of spaceman). His mass is much larger than the tool belt and so his velocity will be smaller, but at least he should now be drifting in the right direction!

1.10.2 Rebounding balls

Another interesting example of momentum change occurs when several balls are dropped onto the ground together (Figure 1.6d). If dropped individually, they rebound to a modest height, less than the height from which they were dropped. But if several balls, each one slightly smaller than the previous one, are now dropped together, one on top of the other, the top one will shoot upwards at an alarming velocity to a height far greater than any of the individual balls. The reason for this is momentum change. The first, larger ball rebounds on impact with the ground and immediately hits the second ball and the second ball hits the third and so on. Each ball transfers its momentum to the next one. As the mass of each ball is smaller, the momentum change increases the velocity. If it was possible to drop eight balls onto each other in this way, it is estimated that the top ball would reach a velocity of 10,000 m/s. This would be fast enough to put it into orbit if it did not vaporise from the heat created by friction as it went through the Earth's atmosphere! Eight balls may be difficult to manage, but even with two or three, the effect is quite dramatic. Try it for yourself.

1.11 PROPERTIES OF WATER

The following physical properties of water will be a useful reference for the later chapters.

1.11.1 Density

When dealing with solid objects, their mass and weight are important, but when dealing with fluids, it is much more useful to know about their *density*. There are two ways of expressing density: *mass density* and *weight density*. Mass density is the mass of one cubic metre of a material and, like mass, it is a fixed value. For example, the mass density of air is 1.29 kg/m³, steel is 7,800 kg/m³ and gold is 19,300 kg/m³.

Mass density is determined by dividing the mass of some object by its volume.

$$\text{mass density (kg/m}^3) = \frac{\text{mass (kg)}}{\text{volume (m}^3)}.$$

Mass density is usually denoted by the Greek letter ρ (rho). One cubic metre of water has a mass of 1,000 kg and so:

$$\rho = 1{,}000 \text{ kg/m}^3.$$

Density can also be written in terms of weight as well as mass. This is referred to as *weight density*, but engineers often use the term *specific weight* (*w*). This is the weight of one cubic metre of water.

Newton's second law is used to link mass and weight:

$$\text{weight density (kN/m}^3) = \text{mass density (kg/m}^3) \times \text{gravity constant (m/s}^2)$$
$$= \rho g.$$

For water:

$$\text{weight density} = 1{,}000 \times 9.81$$
$$= 9{,}810 \text{ N/m}^3 \text{ or } (9.81 \text{ kN/m}^3).$$

Often engineers round-off weight density to 10 kN/m³ for ease of use. This will make very little difference to the calculations for designing hydraulic works. Note that the equation for weight density is applicable to all fluids and not just water. It can be used to find the weight density of any fluid, provided the mass density is known.

Engineers generally use the term specific weight in their calculations, whereas scientists tend to use the term ρg to describe the weight density. Either approach is acceptable, but for clarity, ρg is used throughout this book.

1.11.2 Relative density or specific gravity

Sometimes it is more convenient to use *relative density* rather than just density. It is more commonly referred to as *specific gravity* and is the ratio of the density of a material or fluid to that of some standard density, usually water. It can be written both in terms of the mass density and the weight density.

$$\text{specific gravity (SG)} = \frac{\text{density of an object (kg/m}^3)}{\text{density of water (kg/m}^3)}.$$

Specific gravity has no dimensions. As the volume is 1 m³ for both the object and water, then another way of writing this formula is in terms of weight:

$$\text{specific gravity} = \frac{\text{weight of an object}}{\text{weight of an equal volume of water}}.$$

Some useful specific gravity values are included in Table 1.3.

The density of any other fluid (or any solid object) can be calculated by knowing the specific gravity. The density of mercury, for example, can be calculated from its specific gravity:

$$\text{specific gravity of mercury (SG)} = \frac{\text{density of mercury (kg/m}^3)}{\text{density of water (kg/m}^3)}.$$

Turning this equation around:

$$
\begin{aligned}
\text{mass density of mercury} &= \text{SG of mercury} \times \text{mass density of water} \\
&= 13.6 \times 1{,}000 \\
&= 13{,}600 \text{ kg/m}^3.
\end{aligned}
$$

Table 1.3 Some values of specific gravity

Material/fluid	Specific gravity	Comments
Water	1	All other specific gravity measurements are made relative to that of water
Oil	0.9	Less than 1.0 and so it floats on water
Sand/silt	2.65	Important in sediment transport problems
Mercury	13.6	Fluid used in manometers for measuring pressure

The mass density of mercury is 13.6 times greater than that of water. Archimedes used this idea of specific gravity in his famous principle, which is discussed in Section 2.12.

1.11.3 Viscosity

This is the friction force which exists inside a fluid. It is sometimes referred to as *dynamic viscosity*. To understand this property, imagine a fluid flowing along a pipe as a set of thin layers, each able to slide over each other (Figure 1.7a). The layer nearest to the boundary actually sticks to it and is not moving. The next layer slides over this first layer but is slowed down by friction between them. The third layer moves faster but is slowed by friction with the second layer. This effect continues across the pipe affecting all the flow with layers in the middle of the flow moving fastest. It is similar to a pack of cards when you slide your hand across the pack (Figure 1.7b). The friction between the layers of fluid is known as the *viscosity* and the effect on the flow is referred to as the *boundary layer* which is described in more detail in Section 3.9.3. Some fluids, such as water, have a low viscosity and this means the friction between the layers is low and its influence is not so evident when water is flowing. In contrast, engine oils have a much higher viscosity and they seem to flow more slowly. This is because the viscosity (internal friction) is much greater.

One way to see viscosity at work is to pull out a spoon from a jar of honey. Some of the honey sticks to the spoon and some sticks to the jar, demonstrating that fluid sticks to the boundaries. There is also resistance when you pull out the spoon and this is due to the viscosity. This effect is the same for all fluids, including water, but it is not so obvious as in the honey jar.

Viscosity of water is usually ignored in most hydraulic designs. To take it into account not only complicates the analysis but also has little or no effect on the final design because the forces of viscosity are usually very small relative to other forces involved. When forces of viscosity are ignored, the fluid is described as an *ideal fluid*.

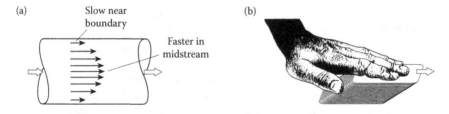

Figure 1.7 Understanding viscosity. (a) Flow in a pipe as a set of thin layers. (b) Flow is similar to a pack of cards.

Another interesting feature which honey demonstrates is that the resistance changes depending on how quickly you pull out the spoon. The faster you pull the spoon, the greater the resistance. Newton found that for most fluids the resistance was proportional to how fast you tried to move the liquid. So resistance increases directly with the velocity of the fluid. Fluids which conform to this rule are known as *Newtonian fluids*.

Some modern fluids have different viscous properties and are called *non-Newtonian fluids*. One good example is tomato ketchup. When left on the shelf, it is a highly viscous fluid which does not flow easily from the bottle. Sometimes you can turn a full bottle upside down and nothing comes out. But shake it vigorously (in scientific terms this means applying a shear force), its viscosity suddenly reduces and the ketchup flows easily from the bottle. In other words, applying a force to a fluid can change its viscous properties often to our advantage.

Although viscosity is often ignored in hydraulics, life would be difficult without it. It would be a world without friction (see Section 1.5). The spoon in the honey jar would come out clean and it would be difficult to get the honey out of the jar. Rivers rely on viscosity to slow down flows otherwise they would continue to accelerate under the influence of gravity to very high speeds. The Mississippi river would reach a speed of over 300 km/h as its flow gradually descends 450 m towards the sea if water had no viscosity. Pumps would not work because impellers would not be able to grip the water and swimmers would not be able to propel themselves through the water for the same reason (see also Box 1.5).

Viscosity is usually denoted by the Greek letter μ (mu).

For water:

$\mu = 0.00114$ kg/ms at a temperature of 15°C

$\quad = 1.14 \times 10^{-3}$ kg/ms.

Viscosity of all Newtonian fluids is influenced by temperature. Viscosity decreases, oils are 'thinner', when temperature increases.

**BOX 1.5   VISCOSITY CAN TELL YOU IF
AN EGG IS HARD OR SOFT BOILED**

A simple test is to spin a boiled egg on the kitchen worktop, stop it suddenly, and let go. If it stays still, then it is hard boiled, if it starts to rotate on its own, it is soft boiled (Figure 1.8).

When the egg is hard boiled, it is solid throughout and so it stays still when you stop it spinning. When is it soft boiled, the liquid inside keeps moving even though you have stopped the egg from spinning. It is the viscous forces which eventually slows the liquid and brings it to a stop.

Figure 1.8 Spin the egg to check if it's soft boiled.

1.11.4 Kinematic viscosity

In many hydraulic calculations, viscosity and mass density go together and so they are often combined into a term known as the *kinematic viscosity*. It is denoted by the Greek letter ν (nu) and is calculated as follows:

$$\text{kinematic viscosity } (\nu) = \frac{\text{viscosity } (\mu)}{\text{density } (\rho)}.$$

For water:

$\nu = 1.14 \times 10^{-6} \text{ m}^2/\text{s}$ at a temperature of 15°C.

Sometimes kinematic viscosity is measured in Stokes in recognition of the work of Sir George Stokes (1819–1903) who helped to develop a fuller understanding of the role of viscosity in fluids.

10^4 Stokes $= 1 \text{ m}^2/\text{s}$.

For water:

$\nu = 1.14 \times 10^{-2}$ Stokes.

1.11.5 Surface tension

An ordinary steel sewing needle can be made to float on water if it is placed there very carefully. A close examination of the water surface around the needle shows that it appears to be sitting in a slight depression and the water behaves as if it is covered with an elastic skin supporting the needle. This fluid property is known as *surface tension*. It is a force but it is very small and is normally expressed in terms of force per unit length across the fluid surface.

For water:

surface tension = 0.51 N/m at a temperature of 20°C.

This force is ignored in hydraulic calculations, but in hydraulic modelling, where small-scale physical models are constructed in a laboratory to work out flow behaviour for large complex in harbours and rivers, surface tension may well influence the behaviour of the model because of the small water depths and flows involved.

1.11.6 Compressibility

It is easy to imagine a gas being compressible and to some extent some solid materials such as rubber. In fact, all materials are compressible; a property which we call *elasticity*. Water is 100 times more compressible than steel! The compressibility of water is important in many aspects of hydraulics. Take for example the task of closing a sluice valve to stop water flowing along a pipeline. If water was incompressible, it would be like trying to stop a 40-ton truck. The water column would be a solid mass running into the valve and the force of impact would be significant. Fortunately, water is compressible, and as it impacts the valve, it compresses like a spring and the energy of the impact is absorbed. But compressing water in this way leads to another problem known as *water hammer* which is discussed more fully in Section 4.14.

Chapter 2

Water standing still

Hydrostatics

2.1 INTRODUCTION

Hydrostatics is the study of water when it is not moving; it is standing still. It is important to civil engineers who are designing water storage tanks and dams. They want to work out the forces that water creates in order to build reservoirs and dams that can resist them. Naval architects designing submarines want to understand and resist the pressures created when they go deep under the sea. The answers come from understanding hydrostatics. The science is simple both in concept and in practice. Indeed, the theory is well established and little has changed since Archimedes (287–212 BC) worked it out over 2000 years ago.

2.2 PRESSURE

The term *pressure* is used to describe the force that water exerts on each square metre of some object submerged in water, that is, force per unit area. It may be the bottom of a tank, the side of a dam, a ship or a submerged submarine. It is calculated as follows:

$$\text{pressure} = \frac{\text{force}}{\text{area}}.$$

Introducing the units of measurement

$$\text{pressure } (\text{kN/m}^2) = \frac{\text{force } (\text{kN})}{\text{area } (\text{m}^2)}.$$

Force is in kilo-Newtons (kN), area is in square metres (m^2) and so pressure is measured in kN/m^2. Sometimes pressure is measured in *Pascals* (Pa) in recognition of Blaise Pascal (1620–1662) who clarified much of modern day thinking about pressure and barometers for measuring atmospheric pressure.

$1\ Pa = 1\ N/m^2$.

One Pascal is a very small quantity and so kilo-Pascals are often used so that

$1\ kPa = 1\ kN/m^2$.

Although it is in order to use Pascals, kN/m^2 is the measure of pressure that engineers tend to use and so this is used throughout this text (see example of calculating pressure in Box 2.1).

2.3 FORCE AND PRESSURE ARE DIFFERENT

Force and pressure are terms that are often confused. The difference between them is best illustrated by an example:

If an elephant or a woman in a high-heel (stiletto) shoe stood on your foot – which is likely to cause the least damage to your foot? We suggest you choose the elephant! To understand this is to appreciate the important difference between force and pressure.

The weight of the elephant is obviously greater than that of the woman but the pressure under the elephant's foot is much less than that under the high-heel shoe (see Box 2.2). The woman's weight (force) is small in comparison to that of the elephant, but the area of the shoe heel is very small and so the pressure is extremely high. So, the high-heel shoe is likely to cause you more pain than the elephant. This is why high-heel shoes, particularly those with a very fine heel, are sometimes banned indoors as they can so easily punch holes in flooring and furniture!

There are many other examples which highlight the difference. Agricultural tractors often use wide (floatation) tyres to spread their load and reduce soil compaction. Military tanks use caterpillar tracks to spread the load to avoid getting bogged down in muddy conditions. Eskimos use shoes shaped like tennis rackets to avoid sinking into the soft snow.

2.4 PRESSURE AND DEPTH

The pressure on some object under water is determined by the depth of water above it. So, the deeper the object is below the surface, the higher will be the pressure. The pressure can be calculated using the pressure-head equation:

$p = \rho g h,$

BOX 2.1   EXAMPLE: CALCULATING PRESSURE

Calculate the pressure on a flat plate 3 m × 2 m when a mass of 50 kg rests on it. Calculate the pressure when the plate is reduced to 1.5 m × 2 m (Figure 2.1).

First, calculate the weight on the plate. Remember weight is a force.

$$\text{mass on plate} = 50 \text{ kg}$$
$$\text{weight on the plate} = \text{mass} \times \text{gravity constant}$$
$$= 50 \times 9.81 = 490.5 \text{ N}$$
$$\text{plate area} = 3 \times 2 = 6 \text{ m}^2$$
$$\text{pressure on plate} = \frac{\text{force}}{\text{area}} = \frac{490.5}{6}$$
$$= 81.75 \text{ N/m}^2.$$

When the plate is reduced to 1.5 m × 2 m

$$\text{plate area} = 1.5 \times 2 = 3 \text{ m}^2$$
$$\text{pressure on plate} = \frac{490.5}{3} = 163.5 \text{ N/m}^2.$$

Note that the mass and the weight remain the same in each case. But the areas of the plate are different and so the pressures are also different.

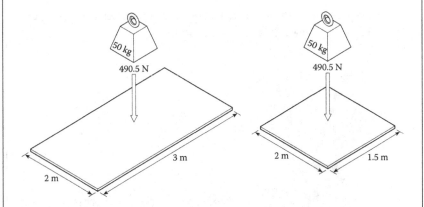

Figure 2.1 Different areas produce different pressures for the same force.

BOX 2.2   EXAMPLE: THE ELEPHANT'S FOOT AND THE WOMAN'S SHOE

An elephant has a mass of 5,000 kg and its feet are 0.3 m in diameter. A woman has a mass of 60 kg and her shoe heel has a diameter of 0.01 m. Which produces the greater pressure – the elephant's foot or woman's shoe heel (Figure 2.2)?

First, calculate the pressure under the elephant's foot

$$\text{elephant's mass} = 5,000 \text{ kg}$$

$$\text{elephant's weight} = 5,000 \times 9.81 = 49 \text{ kN}$$

$$\text{weight on each foot} = \frac{49}{4} = 12.25 \text{ kN}$$

$$\text{foot area} = \frac{\pi d^2}{4} = \frac{\pi 0.3^2}{4} = 0.07 \text{ m}^2$$

$$\text{pressure under foot} = \frac{\text{force}}{\text{area}} = \frac{12.25}{0.07} = 175 \text{ kN/m}^2.$$

Now, calculate the pressure under the woman's shoe heel

$$\text{woman's mass} = 60 \text{ kg}$$

$$\text{woman's weight} = 589 \text{ N} = 0.59 \text{ kN}$$

$$\text{weight on each foot} = \frac{0.59}{2} = 0.29 \text{ kN}$$

$$\text{area of shoe heel} = \frac{\pi d^2}{4} = \frac{\pi 0.01^2}{4} = 0.0001 \text{ m}^2$$

$$\text{pressure under heel} = \frac{\text{force}}{\text{area}} = \frac{0.29}{0.0001} = 2,900 \text{ kN/m}^2.$$

The pressure under the woman's heel is 16 times greater than under the elephant's foot! So beware who treads on your toes.

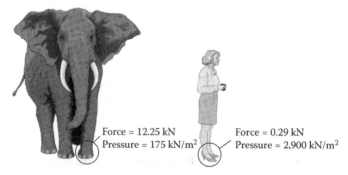

Force = 12.25 kN
Pressure = 175 kN/m²

Force = 0.29 kN
Pressure = 2,900 kN/m²

Figure 2.2 Which produces the greater pressure?

where p is the pressure (N/m²), ρ the mass density of water (kg/m³), g the gravity constant (m/s²) and h the depth of water (m).

This equation works for all fluids and not just water, provided of course that the correct value of density is used for the fluid concerned.

If you would like to see how the pressure-head equation is derived look in Box 2.3.

BOX 2.3   DERIVATION: PRESSURE-HEAD EQUATION

Imagine a tank of water of depth h and a cross-sectional area a (Figure 2.3). The weight of water on the bottom of the tank (remember that weight is a force and is acting downwards) is balanced by an upward force from the bottom of the tank supporting the water (Newton's third law). The pressure-head equation is derived by calculating these two forces and putting them equal to each other.

First, calculate the downward force. This is the weight of water. To do this, first calculate the volume and then the weight using the density

volume of water = cross sectional area × depth.

$$= a \times h$$

And so

weight of water in tank = volume × density × gravity constant.

$$= a \times h \times \rho \times g$$

This is the downward force of the water ↓. Next, calculate the supporting (upward) force from the base

supporting force = pressure × area.

$$= p \times a$$

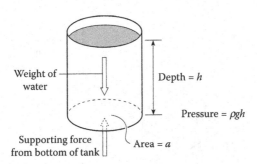

Figure 2.3 Calculating the force on the bottom of a tank.

Now, put these two forces equal to each other

$$p \times a = a \times h \times \rho \times g,$$

The area a cancels out from both sides of the equation and so we are left with

$$p = \rho g h.$$

In words

pressure (N/m²) = mass density (kg/m³) × gravity constant (9.81 m/s²)
× depth of water (m).

This is the *pressure-head equation* which links pressure with the depth of water. It shows that pressure increases directly as the depth increases. Note that it is completely independent of the shape of the tank or the area of base.

BOX 2.4   EXAMPLE: CALCULATING PRESSURE AND FORCE ON THE BASE OF A WATER TANK

A rectangular tank of water is 3 m deep. If the base measures 3 m by 2 m, calculate the pressure and force on the base of the tank (Figure 2.4).

Use the pressure-head equation

$$p = \rho g h$$
$$= 1,000 \times 9.81 \times 3.0 = 29,430 \, \text{N/m}^2 = 29.43 \, \text{kN/m}^2.$$

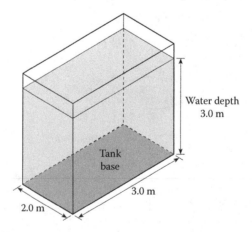

Water depth
3.0 m

Tank base

3.0 m

2.0 m

Figure 2.4 Calculating force and pressure on the base of a tank.

Calculate the force on the tank base using the pressure and the area

$$\text{force} = \text{pressure} \times \text{area}$$
$$\text{base area} = 3 \times 2 = 6\,\text{m}^2$$
$$\text{force} = 29.43 \times 6 = 176.6\,\text{kN}.$$

2.5 PRESSURE IS SAME IN ALL DIRECTIONS

Although in the example in Box 2.4, the pressure is used to calculate the downward force on the tank base, pressure does not in fact have a specific direction – it pushes in all directions. To understand this, imagine a cube in the water (Figure 2.5). The water pressure pushes on all sides of the cube and not just on the top. If the cube was very small, then the pressure on all six faces would be almost the same. If the cube gets smaller and smaller until it almost disappears to a point, it becomes clear that *the pressure at a point in the water is the same in all directions*. So, the pressure pushes in all directions and not just vertically. This idea is important for designing dams because it is the horizontal action of pressure which pushes on a dam and which must be resisted if the dam is not to fail. Note also that the 'pressures' in Figure 2.5 are drawn pushing inward. But they could equally have been drawn pushing outward to make the same argument – remember Newton's third law.

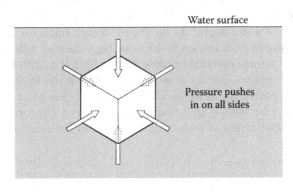

Figure 2.5 Pressure is the same in all directions.

2.6 HYDROSTATIC PARADOX

It is often assumed that the size of a water tank or its shape influences pressure but this is not the case (Figure 2.6). It does not matter if the water is in a large tank or in a narrow tube. The pressure-head equation tells us that water depth is the only variable that determines the pressure. So, the base area neither has effect on the pressure nor does the amount of water in the tank. What is different of course is the force on the base of different containers. The *force* on the base of each tank is equal to the weight of water in each of the containers. But if the depth of water in each is the same then the pressure will always be the same.

2.6.1 Bucket problem

The Dutch mathematician Simon Stevin (1548–1620) made a similar point by showing that the *force* on the base of a tank depended only on the area of the base and the vertical depth of water (the pressure) – and not the weight of water in the container. This is well demonstrated using three different shaped buckets but each with the same base area and the same depth of water in them (Figure 2.7).

The weight of water in each bucket is clearly different and a casual observer might assume from this that the force on the base of each bucket will also be different; the force on the base of bucket *b* being greater than bucket *a* and the force on bucket *c* being less than in bucket *a*. But thinking about this in hydraulic terms, the pressure-head equation tells us otherwise. In fact, it predicts that the force on the base is the *same* in each bucket, equal to the area of the base multiplied by the pressure on the base, which is only a function of the water depth. As the depth in each bucket is the same, the force on the base of each bucket is the same regardless of whether the sides are vertical or inclined inwards or outwards. In bucket *a*, with vertical sides, the force does, in fact, equal the weight of water in the bucket. But when the sides slope outwards, as in bucket *b*, the force is less than the weight of water and when the sides slope inwards, as in bucket *c*, the force on the base is greater than the weight of water. All this seems rather absurd but it is true.

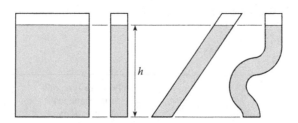

Figure 2.6 Pressure is the same at the base of all the containers.

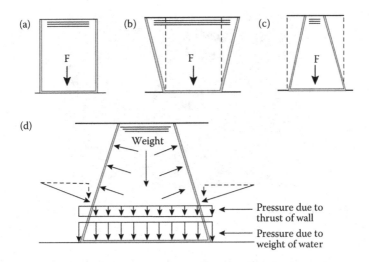

Figure 2.7 Bucket problem.

The key to the paradox lies with the fact that the pressure at a point in the water is the same in all directions (see Section 2.5). The water not only pushes down onto the base but also pushes on the sides of the container as well. So, when the sides slope inwards (bucket *c*), the water pushes outwards and also upwards. Newton's second law says that this produces a corresponding downward force on the water and this is transmitted to the base adding to the force due to the weight of the water. In fact, the total vertical force on the walls and base (the force on the base minus the upward force on the walls) is exactly equal to the weight of water in the bucket! The same argument can be applied to bucket *b*. The water pushes on the sides of the tank and in this case pushes outwards and downwards. Newton's second law says that this produces a corresponding upward force on the water and this is transmitted to the base reducing the force due to the weight of the water. So, in this case the force on the base is less than the weight of the water in the bucket.

Clear? If so then you are well on your way not only to understand the important difference between force and pressure but also to appreciate the significance of Newton's contribution to our understanding of the way in which our water world works.

2.6.2 Balloon problem

One more 'absurdity' to test your understanding. Two identical balloons are connected to a manifold and blown up independently so that one is larger than the other (Figure 2.8). When valve 3 is closed and valves 1 and

Figure 2.8 Balloon problem.

2 are opened, the air can flow between the balloons to equalise the air pressure. The question is – What happens to the balloons?

You might expect the air to flow from the larger balloon to the smaller one, so they both become the same size. But this is not what happens. The larger balloon, in fact, gets larger and the smaller balloon gets smaller. The reason for this is again explained by the difference between pressure and force. The larger balloon has a much greater surface area than the smaller one and so the pressure inside the larger balloon is less than in the smaller balloon. So, when the two balloons are connected together by opening the valves 1 and 2, the air flows from the smaller balloon into the larger balloon thus making the larger balloon even larger and the smaller balloon even smaller.

Do not confuse size with pressure. If you are not convinced or you are still confused, try the balloon experiment by making up a small manifold using some plastic pipes and laboratory taps, and see for yourself.

2.7 PRESSURE HEAD

Engineers often refer to pressure in terms of metres of water rather than as a pressure in kN/m^2. Referring back to the pressure calculation in Box 2.4, instead of saying the pressure is 29.43 kN/m^2, we can say the pressure is '3 m head of water'. This is because of the unique relationship between the pressure and the water depth ($p = \rho g h$). It is called the *pressure head* or just *head* and is measured in metres. It is the water depth h referred to in the pressure-head equation. Both ways of stating the pressure are correct and one can easily be converted to the other using pressure-head equation.

**BOX 2.5   EXAMPLE: CALCULATING
PRESSURE HEAD IN MERCURY**

Building on the example in Box 2.4, calculate the depth of mercury needed in the tank to produce the same pressure as 3 m depth of water (29.43 kN/m^2). Specific gravity (SG) of mercury is 13.6.

First, calculate the density of mercury

$$\rho \text{ (mercury)} = \rho \text{ (water)} \times \text{SG (mercury)}$$
$$= 1,000 \times 13.6$$
$$= 13,600 \text{ kg/m}^3.$$

Use pressure-head equation to calculate the head of mercury

$$p = \rho g h,$$

where ρ is now the density of mercury and h is the depth

$$29,430 = 13,600 \times 9.81 \times h$$
$$h = 0.22 \text{ m of mercury.}$$

So, the depth of mercury required to create the same pressure as 3 m of water is only 0.22 m. This is because mercury is much denser than water.

Engineers prefer to use head measurements because, as will be seen later, differences in ground level can affect the pressure in a pipeline. It is an easy matter to add (or subtract) changes in ground level to pressure values because they both have the same units.

A word of warning though. When head is measured in metres it is important to say what the liquid is – '3 m head of water' will be a very different pressure to '3 m head of mercury'. This is because the density term ρ is different. So, the rule is – say what liquid is being used to measure the pressure. See the worked example in Box 2.5.

2.8 ATMOSPHERIC PRESSURE

The pressure of the atmosphere is all around us pressing on our bodies. Although, we often talk about things being 'as light as air' when there is a large depth of air, it creates a high pressure of approximately 100 kN/m^2 on the Earth's surface – the equivalent of 10 m head of water. The average person has a skin area of 2 m², and so the force acting on each of us from the air around us is approximately 200 kN (the equivalent of 200,000 apples or 20 tons). A very large force indeed! Fortunately, there is an equal and opposite pressure from within our bodies that balances the air pressure and so we feel no effect (Newton's third law).

At high altitudes, atmospheric pressure is less than at the Earth's surface, and some people suffer from nose bleeds due to their blood pressure being much higher than the surrounding atmosphere. We can also detect sudden changes in air pressure. For instance, when we fly in an aeroplane, even though the cabin is pressurised, our ears pop as our bodies adjust to changes in the cabin pressure. If the cabin pressure system failed suddenly, removing one side of this pressure balance then the result could be catastrophic. Inert gases such as nitrogen, which are normally dissolved in our body fluids and tissues, would rapidly start to form gas bubbles which can result in sensory failure, paralysis and death. Deep sea divers are well aware of this rapid pressure change problem, known as the bends, and they make sure that they return to the surface slowly so that their bodies have enough time to adjust to the changing pressure. A good practical demonstration of what happens can be seen when you open a fizzy drink bottle. When the cap is removed from the bottle, gas is heard escaping, and bubbles can be seen forming in the drink. This is carbon dioxide gas coming out of the solution as a result of the sudden pressure drop inside the bottle as it equalises with the pressure of the atmosphere.

It was in the seventeenth century that scientists, such as Evangelista Torricelli (1608–1647), a pupil of Galileo Galilei (1564–1642), began to understand about atmospheric pressure and to study the importance of vacuums – the empty space when all the air is removed. Scientists previously explained atmospheric effects by saying *that nature abhors a vacuum*. This meant that if the air is sucked out of a bottle, it will immediately fill by sucking air back in again when it is opened to the atmosphere. But Galileo had already observed that a suction pump could not lift water more than 10 m, so there appeared to be a limit to this abhorrence. Today, we know that it is not the vacuum in the bottle that sucks in the air but the outside air pressure that pushes the air in to fill the vacuum. The end result is the same (i.e. the bottle is filled with air) but the mechanism is quite different and has implications for suction pumps.

Suction pumps do not 'suck' up water as is commonly thought. It is the atmospheric pressure on the surface of the water that pushes water into a pump when the air inside is removed to create a vacuum – a process known as 'priming'. The implication of this is that the atmospheric pressure (which is 10 m head of water) puts a limit on how high a pump can be located above the water surface and still work. In practice, the limit is a lot lower than this, but more about this is given in Section 7.4. Siphons too rely on atmospheric pressure in a similar way (Section 4.7).

Atmospheric pressure does vary over the Earth's surface. It is lower in mountainous regions and also varies as a result of the Earth's rotation and temperature changes in the atmosphere, both of which cause large movements of air. They create high- and low-pressure areas and winds as air flows from high-pressure to low-pressure areas in an attempt to try and equalise the air pressure. This is an important phenomenon for the world's weather but in hydraulics such differences are relatively small and have little effect on solving problems – except of course if you happen to be building a pumping station for a community in the Andes or the Alps. So for all intents and purposes, atmospheric pressure close to sea level can be assumed constant at 100 kN/m^2 – or 10 m head of water. For example experiencing atmospheric pressure, see Box 2.6.

BOX 2.6   EXAMPLE: EXPERIENCING ATMOSPHERIC PRESSURE

One way of experiencing the effect of atmospheric pressure is to place a large sheet of paper on a table over a thin piece of wood (Figure 2.9). If you hit the wood sharply it is possible to strike a considerable blow without disturbing the paper. You may even break the wood. This is because the paper is being held down by the pressure of the atmosphere.

If the paper is 1.0 m^2, the force holding down the paper can be calculated as follows:

force = pressure × area.

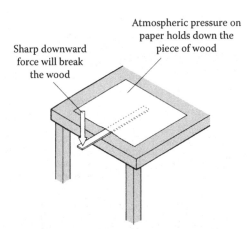

Atmospheric pressure on paper holds down the piece of wood

Sharp downward force will break the wood

Figure 2.9 Experiencing atmospheric pressure.

In this case

pressure = atmospheric pressure = 100 kN/m².

And so

force holding the paper down = 100 × 1 = 100 kN.

In terms of apples this is about 100,000 or 10 tons, which is a large force. It is little wonder that the wood breaks before the paper lifts.

2.8.1 Mercury barometer

An instrument for measuring atmospheric pressure is the mercury barometer. Evangelista Torricelli, is credited with inventing this device in 1643 which is still very much in use today. The only modification was added by Jean Fortin (1750–1831), an instrument maker, who introduced a vernier measuring scale so that minor changes in atmospheric pressure could be accurately measured. Hence, the common name of *Fortin barometer*.

Torricelli's barometer consists of a vertical glass tube closed at one end, filled with mercury and inverted with the open end immersed in a tank of mercury which creates a vacuum in the top of the tube (Figure 2.10). The mercury surface in the tank is exposed to atmospheric pressure and this

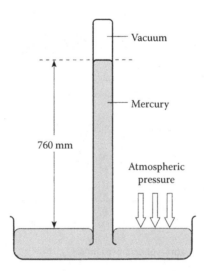

Figure 2.10 Measuring atmospheric pressure using a mercury barometer.

BOX 2.7   EXAMPLE: MEASURING ATMOSPHERIC PRESSURE USING A MERCURY BAROMETER

Calculate atmospheric pressure when the reading on a mercury barometer is 760 mm of mercury (Figure 2.10). What would be the height of the column if the same air pressure was measured using water instead of mercury?

The pressure-head equation links together atmospheric pressure and the height of the mercury column but remember the fluid is now mercury and not water

atmospheric pressure $= \rho gh$,

where h is 760 mm and ρ for mercury is 13,600 kg/m³ (13.6 times denser than water).

So,

atmospheric pressure $= 13.6 \times 9.81 \times 0.76$

$= 101{,}400$ N/m² or 101.4 kN/m².

Calculate the height of the water column to measure the same atmospheric pressure using the pressure-head equation again

atmospheric pressure $= \rho gh$.

This time the fluid is water and so

$101{,}400 = 1{,}000 \times 9.81 \times h$

$h = 10.32$ m.

This is a very tall water column and there would be practical difficulties if this was used for routine measurement of atmospheric pressure. Hence, the reason why a very dense liquid-like mercury is used to make measurement more manageable.

supports the mercury column, the height of which is a measure of atmospheric pressure. A typical value would be 760 mm when it is measured at sea level (see example in Box 2.7).

Torricelli used mercury instead of water because it is significantly denser and makes a more manageable measuring device. If he had used water, he would have needed a tube over 10 m high to do it – not a very practical proposition for use in a laboratory or for taking measurements.

BOX 2.8   EXAMPLE: CALCULATING PRESSURE HEAD

A pipeline is operating at a pressure of 3.5 bar. Calculate the pressure in metres head of water.

I bar $= 100$ kN/m^2 $= 100,000$ N/m^2.

And so

3.5 bar $= 350$ kN/m^2 $= 350,000$ N/m^2.

Use the pressure-head equation

$$p = \rho g h$$
$$350,000 = 1,000 \times 9.81 \times h.$$

Calculate head h

$h = 35.67$ m.

Round this off: 3.5 bar $= 36$ m of head water (approximately).

Atmospheric pressure is also used as a unit of measurement for pressure both for meteorological purposes and in hydraulics. This is known as the *bar* (see example in Box 2.8). For convenience, 1 bar pressure is rounded off to 100 kN/m^2.

A more commonly used term in meteorology is the *millibar*. So,

1 millibar $= 0.1$ kN/m^2 $= 100$ N/m^2.

To summarise – there are several ways of expressing atmospheric pressure

atmospheric pressure $= 1$ bar

or $= 100$ kN/m^2

or $= 10$ m head of water

or $= 760$ mm head of mercury.

2.9 MEASURING PRESSURE

2.9.1 Gauge and absolute pressures

Pressure measuring devices work in the atmosphere with normal atmospheric pressure all around them. Rather than adding atmospheric pressure

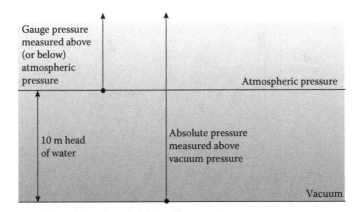

Figure 2.11 Measuring gauge and absolute pressures.

each time a measurement is made, it is a common practice to assume that the atmospheric pressure is equal to zero and so it becomes the base line (or zero point) from which all pressure measurements are made. It is rather like setting sea level as zero from which all ground elevations are measured (Figure 2.11). Pressures measured in this way are called *gauge pressures*. They can be either positive (above atmospheric pressure) or negative (below atmospheric pressure).

Most pressure measurements in civil engineering are gauge pressures but some mechanical engineers, working with gas systems occasionally measure pressure using a vacuum as the datum. In such cases the pressure is referred to as *absolute pressure*. The vacuum is now the zero pressure. As it is not possible to go below vacuum, all absolute pressures have positive values.

To summarise:

Gauge pressures are pressures measured above or below atmospheric pressure. Absolute pressures are pressures measured above a vacuum.

To change from one to the other

absolute pressure = gauge pressure + atmospheric pressure.

Note, if only the word pressure is used, it is reasonable to assume that this means gauge pressure.

2.9.2 Bourdon gauges

Pressure can be measured in several ways. The most common instrument is the *Bourdon gauge* (Figure 2.12a). This is located at some convenient point on a pipeline or pump to record pressure, usually in kN/m^2 or bar pressure. It is a simple device and works on the same principle as a party toy that uncoils when you blow into it. Inside a Bourdon gauge, there is a similar

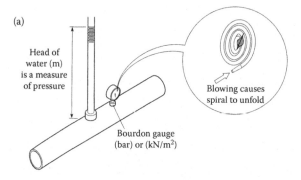

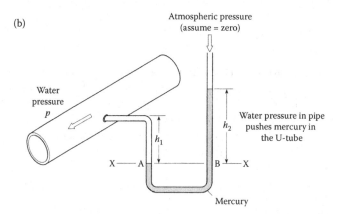

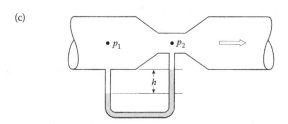

Figure 2.12 Ways of measuring pressure. (a) Bourdon gauge and piezometer. (b) U-tube manometer. (c) Venturi flow meter with manometer.

curved tube which tries to straighten out under pressure and causes a pointer to move through a gearing system across a scale of pressure values.

2.9.3 Piezometers

This is another device for measuring pressure. A vertical tube is connected to a pipe so that water can rise up the tube because of the pressure in the

---

BOX 2.9   EXAMPLE: MEASURING PRESSURE USING A STANDPIPE

Calculate the height of a standpipe needed to measure a pressure of 200 kN/m² in a water pipe (Figure 2.12a).

Use the pressure-head equation

$$p = \rho gh$$
$$200,000 = 1,000 \times 9.81 \times h.$$

Note in the equation pressure and density are both in N – not kN. Rearranging this equation to calculate h.

$$h = 20.4 \text{ m}.$$

A very high tube would be needed to measure this pressure and it would be a rather impracticable measuring device! For this reason high pressures are normally measured using a Bourdon gauge.

---

pipe (Figure 2.12a). This is called a *piezometer* or *standpipe*. The height of the water column in the tube is a measure of the pressure in the pipe, that is, the pressure head. The pressure in kN/m² can be calculated using the pressure-head equation (see Box 2.9).

2.9.4 Manometers

Vertical standpipes are not very practical for measuring high pressures. An alternative is to use a *U-tube manometer* (Figure 2.12b).

The bottom of the U-tube is filled with a different liquid which does not mix with that in the pipe. When measuring pressures in a water system, oil or mercury is used. Mercury is very useful because high pressures can be measured with a relatively small tube (see atmospheric pressure).

To measure pressure, a manometer is connected to a pipeline and mercury is placed in the bottom of the U bend. The basic assumption is that as the mercury in the manometer is not moving, the pressures in the two limbs must be the same. If a horizontal line x–x is drawn through the mercury surface in the first limb and extended to the second limb then it can be assumed that

pressure at point A = pressure at point B.

This is the fundamental assumption on which all manometer calculations are based. It is then a matter of adding up all the components which make up the pressures at A and B to work out a value for the pressure in the pipe.

First, calculate the pressure at A

pressure at A = water pressure at centre of pipe (p)

+ pressure due to water column h_1

$= p + \rho_{(water)}gh_1$

$= p + (1,000 \times 9.81 \times h_1)$

$= p + (9,810 \times h_1)$.

Now, calculate the pressure at B

pressure at B = pressure due to mercury column h_2

+ atmospheric pressure.

As we are measuring gauge pressure, atmospheric pressure is zero. So,

pressure at B = $\rho_{(mercury)}\, gh_2$

$= 1,000 \times 13.6 \times 9.81 \times h_2$

$= 133,430h_2$.

Putting the pressure at A equals to the pressure at B

$p + 9,810h_1 = 133,430h_2$.

Rearrange this to determine the pressure p in the pipe

$p = 133,430h_2 - 9,810h_1$.

Note that pressure p is in N/m².

So, the pressure in this pipeline can be calculated by measuring h_1 and h_2 and using the above equation (see example in Box 2.10).

Some manometers are used to measure pressure differences rather than actual values of pressure. One example of this is the measurement of the pressure difference in a venturi meter used to measure pipe discharge (Figure 2.12c). There is a drop (difference) in pressure as water flows through the narrow venturi. By connecting one limb of the manometer to the main pipe and the other limb to the venturi, the difference in pressure can be measured. Note that the pressure difference is not just the difference in the mercury readings on the two columns as is often thought. The pressure difference must be calculated using the principle described above for the simple manometer.

The best way to deal with manometer measurements is to remember the principle on which all manometer calculations are based and not the

---

BOX 2.10   EXAMPLE: MEASURING PRESSURE USING A MANOMETER

A mercury manometer is used to measure the pressure in a water pipe (Figure 2.12b). Calculate the pressure in the pipe when $h_1 = 1.5$ m and $h_2 = 0.8$ m.

To solve this problem, start with the principle on which all manometers are based

pressure at A = pressure at B

pressure at A = water pressure in pipe (p)

$\qquad$ + pressure due to water column h_1

$\qquad = p + \rho_{(water)}\, gh_1$

$\qquad = p + 1,000 \times 9.81 \times 1.5$

pressure at B = pressure due to mercury column h_2

$\qquad$ + atmospheric pressure (=zero)

$\qquad = \rho_{(mercury)}\, gh_2$

$\qquad = 1,000 \times 13.6 \times 9.81 \times 0.8.$

Putting the pressure at A equal to the pressure at B

$p + 1,000 \times 9.81 \times 1.5 = 1,000 \times 13.6 \times 9.81 \times 0.8.$

Rearrange this to determine p

$p = (1,000 \times 13.6 \times 9.81 \times 0.8) - (1,000 \times 9.81 \times 1.5)$

$\quad = 106,732 - 14,715$

$\quad = 92,017\ \text{N/m}^2 = 92\ \text{kN/m}^2.$

---

formula for pressure. There are many different ways of arranging manometers with different fluids in them and so there will be too many formulae to remember. So, just remember and apply the principle – pressure on each side of the manometer is the same across a horizontal line A–B.

2.10 DESIGNING DAMS

Engineers are always interested in the ways in which things fall down or collapse so they can devise design and construction procedures that produce safe reliable structures. Dams in particular are critical structures because failure can cause a great deal of damage and loss of life. Hydraulically, a

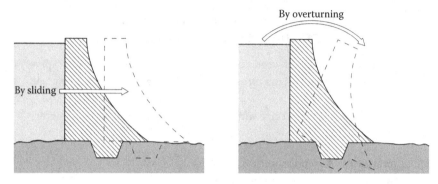

Figure 2.13 Dams can fail by sliding and overturning.

dam structure can fail in two ways – the pressure of water can cause the dam to slide forward and it can also cause it to overturn (Figure 2.13). The engineer must design a structure that is strong enough to resist both these possible modes of failure. This is where the principles of hydrostatics play a key role – the same principles apply to small dams only a few metres high and to major dams 40 m or more in height.

The pressure of water stored behind a dam produces a horizontal force which could cause it to slide forward if the dam was not strong enough to resist. So, the total force resulting from the water pressure must first be calculated. The location of this force is also important. If it is near the top of the dam, it may cause the dam to overturn. If it is near the base, then it may fail by sliding.

The force on a dam is calculated from the water pressure (Figure 2.14a). Remember that pressure pushes in all directions and in this case it is the horizontal push on the dam which is important. At the water surface the pressure is zero, but 1.0 m below the surface the pressure rises to 10 kN/m^2 (approximately), at 2.0 m it reaches 20 kN/m, and so on (remember the pressure-head equation $p = \rho g h$). A graph of the changes in pressure with depth helps to visualise what is happening. It is triangular in shape and is called the *pressure diagram*. It not only shows how pressure varies with depth on the upstream face of a dam but is also very useful for calculating the force on the dam and its location.

The force on the dam can be calculated from the pressure and the area of the dam face using the equation $F = pa$. But the pressure is not constant – it varies down the face of the dam and so the question is – which value of pressure to use? One approach is to divide the dam face into lots of small areas and use the average pressure for each small area to calculate the force at that point. All the forces are then added together to find the total force on the dam. But this is a rather tedious approach. A much simpler method is to use a formula derived from combining all the small forces mathematically into a single large force known as the resultant force F (Figure 2.14a).

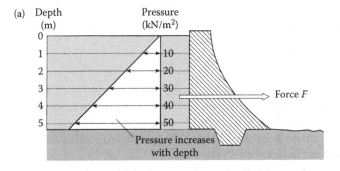

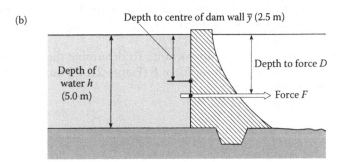

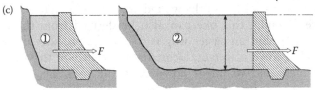

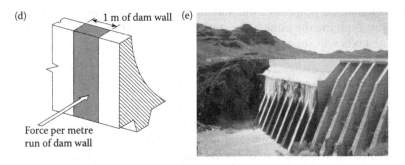

Figure 2.14 Designing a dam. (a) Pressure diagram. (b) Location of force. (c) Dam paradox. (d) Assume dam is 1.0 m long. (e) Typical concrete dam.

This single force has the same effect as the sum of all the smaller forces and is much easier to deal with. The formula for F is

$$\text{force } (F) = \rho g a \bar{y},$$

where ρ is the density of water (kg/m³), g the gravity constant (9.81 m/s²), a the area of the wetted face of the dam (m²) and $\bar{y}$ the depth from the water surface to centre of the wetted area (m).

For the mathematically minded a derivation of this formula can be found in most engineering hydraulics textbooks.

We can also work out the force from the pressure diagram. It is in fact, equal to the area of the diagram, that is, the area of the triangle. To see how this, and the formula for force, works look at the example of how to calculate the force on a dam in Box 2.11.

The position of this force is also important. To determine the depth D from the water surface to the resultant force F (Figure 2.14b) on the dam, the following formula can be used:

$$D = \frac{h^2}{12\bar{y}} + \bar{y},$$

where h is the height of the wetted face of the dam (m) and $\bar{y}$ the depth from water surface to the centre of the wetted area of the dam (m).

Like the force formula, this one can also be derived from the principles of hydrostatics. But an easier way is to use the pressure diagram to determine D. The force is in fact, located at the centre of the diagram which, for a triangle, is two thirds down from the apex (i.e. from the water surface).

Note that these formulae only work for simple vertical dams. When more complex shapes are involved, such as earth dams with sloping sides, then the formulae do not work. But solving the problem is not so difficult – it relies on applying the same hydrostatic principles. Most standard civil engineering texts will show you how.

2.10.1 Dam paradox

Dams raise an interesting paradox. If two dams are built and are the same height but hold back very different amounts of water how will they differ in their hydraulic design (Figure 2.14c)? Many would say that dam 2 would need to be much stronger than dam 1 because it is holding back more water. But this is not so. Hydraulically, the design of the two dams will be the same. The force on dam 1 is the same as on dam 2 because the force depends only on the depth of water and not the amount stored. The effects of failure would obviously be more serious with dam 2 as the

BOX 2.11   EXAMPLE: CALCULATING THE FORCE ON A DAM

A farm dam is to be constructed to contain water up to 5 m deep. Calculate the force on the dam and the position of the force in relation to the water surface (Figure 2.14b).

Calculate the force F

$$F = \rho g a \bar{y},$$

where $\rho = 1,000$ kg/m³, $g = 9.81$ m/s² and $a = b \times h = 1 \times 5 = 5$ m.

Note: When the length of the dam is not given assume that $b = 1$ m. The force is then the force per metre length of the dam (Figure 2.14d).

$$\bar{y} = \frac{h}{2} = 2.5 \text{ m.}$$

Put all these values into the formula for F

$$F = 1,000 \times 9.81 \times 5 \times 2.5$$
$$= 122,625 \text{ N}$$
$$F = 122.6 \text{ kN per m length of the dam.}$$

Using the alternative method of calculating the area of the pressure diagram

$$\text{area of pressure diagram (triangle)} = \frac{1}{2} \times \text{base} \times \text{height}$$
$$= \frac{1}{2} \times \rho g h \times h$$
$$= \frac{1}{2} \times 1,000 \times 9.81 \times 5 \times 5$$
$$F = 122.6 \text{ kN per m length of the dam.}$$

This produces the same answer as the formula.
To locate the force use the formula

$$D = \frac{h^2}{12\bar{y}} + \bar{y}$$
$$= \frac{5^2}{12 \times 2.5} + 2.5$$
$$D = 3.33 \text{ m below the water surface.}$$

Using the pressure diagram method, the force is located at the centre of the triangle, which is two-thirds down from the water surface

$$D = \frac{2}{3} \times h = \frac{2}{3} \times 5$$

$D = 3.33$ m below the water surface.

This produces the same answer as the formula.

potential for damage and loss of life from all that extra water could be immense. So, the designer may introduce extra factors of safety against failure. So if you thought the forces would be different, place your trust in the well-established principles of hydrostatics and not your intuition.

2.11 FORCES ON SLUICE GATES

Sluice gates are used to control the flow of water from dams into pipes and channels. They may be circular or rectangular in shape and are raised and lowered by turning a wheel on a threaded shaft (Figure 2.15a).

Gates must be made strong enough to withstand the forces created by hydrostatic pressure. The pressure also forces the gate against the face of the dam which can make it difficult to lift easily because of the friction it creates. So, the greater the pressure the greater will be the force required to lift the gate. This is the reason why some gates have gears and hand wheels fitted to make lifting easier.

The force on a gate and its location can be calculated in the same way as for a dam. The force on any gate can be calculated using the same formula as was used for the dam

$$F = \rho g a \bar{y}.$$

In this case, a is the area of the gate and $\bar{y}$ the depth from the water surface to the centre of the gate. The formula for calculating D, the depth to the force, depends on the shape of the gate.

For rectangular gates

$$D = \frac{d^2}{12\bar{y}} + \bar{y},$$

where d is the depth of the gate (m) and $\bar{y}$ the depth from the water surface to the centre of the gate (m). Note: in this case d is the depth of the gate (m) and *not* the depth of water behind the dam.

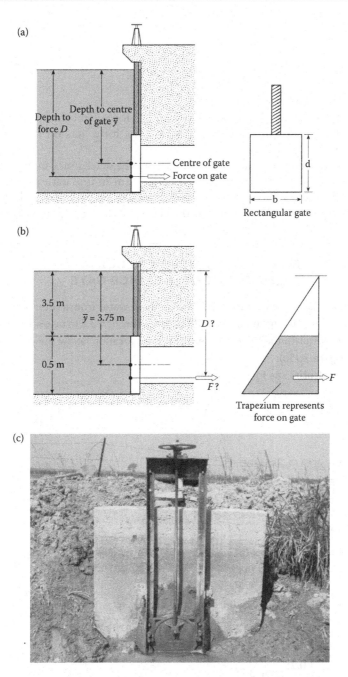

Figure 2.15 Forces on sluice gates. (a) Force on rectangular sluice gate. (b) Pressure diagram. (c) Typical circular sluice gate.

For circular gates

$$D = \frac{r^2}{4\bar{y}} + \bar{y},$$

where r is the radius of the gate (m).

The depth D from the water surface to the force F must not be confused with $\bar{y}$. D is the depth to the point where the force acts on the gate. It is always greater than $\bar{y}$.

The force and its location on a gate can also be determined using the pressure diagram, but only that part of the diagram in line with the gate is of interest (Figure 2.15b). The force on the gate is calculated from the area of the trapezium and its location is at the centre of the trapezium. This can be

---

BOX 2.12 EXAMPLE: CALCULATING THE FORCE ON A SLUICE GATE

A rectangular sluice gate controls the release of water from a reservoir. If the gate is 0.5 m × 0.5 m and the top of the gate is located 3.5 m below the water surface, calculate the force on the gate and its location below the water surface (Figure 2.15b).

First calculate the force F on the gate

$$F = \rho g a \bar{y},$$

where

a = area of the gate = 0.5 × 0.5 = 0.25 m^2
$\bar{y}$ = depth from water surface to the centre of the gate
   = 3.5 + 0.25 = 3.75 m.

And so

$F = 1{,}000 \times 9.81 \times 0.25 \times 3.75$
$F = 9{,}120$ N or 9.12 kN.

Next, calculate the depth from water surface to where force F is acting

$$D = \frac{d^2}{12\bar{y}} + \bar{y}$$

$$= \frac{0.25}{12 \times 3.75} + 3.75$$

$D = 3.76$ m.

found by using the principle of moments. But if you are not so familiar with moments, the centre can be found by cutting out a paper shape of the trapezium and freely suspending it from each corner in turn and drawing a vertical line across the shape. The point where all the lines cross is the centre. A common mistake is to assume the depth D is two thirds of the depth from the water surface. It is true for a simple dam but not for a sluice gate.

The above equations cover most hydraulic sluice gate problems but occasionally gates come in different shapes and are sometimes at an angle rather than vertical. It is still possible to work out the forces on such gates but it is a bit more involved. Other hydraulic textbooks will show you how, if you are curious enough. An example of calculating the force and its location on a hydraulic gate is shown in Box 2.12.

2.12 ARCHIMEDES PRINCIPLE

Returning now to Archimedes who first set down the basic rules of hydrostatics. His most famous venture seems to have been in the public baths in Greece around the year 250 BC. He allegedly ran naked into the street shouting *eureka* – he had discovered an experimental method of detecting the gold content of the crown of the King of Syracuse. He realised that when he got into his bath, the water level rose around him because his body was displacing the water and that this was linked to the feeling of weight loss; that uplifting feeling that everyone experiences in the bath. As the baths were usually public places, he probably noticed as well that smaller people displaced less water. It is at this point that many people draw the wrong conclusion. They assume that this has something to do with a person's weight. This is quite wrong – it is all about their volume. To explain this, let us return to the king's crown.

Perhaps the king had two crowns that looked the same in every way, but one was made of gold and he suspected that someone had short-changed him by making the other of a mixture of gold and some cheaper metal. The problem that he set to Archimedes was to tell him which was the gold one. Weighing them on a normal balance in air would not have provided the answer because a clever forger would make sure that both crowns were the same weight. If however, he could measure their densities he would then know which was gold because the density of gold has a fixed value (19,300 kg/m^3) and this would be different to the crown of mixed metals. But to determine their densities you need to be able to measure the crown volumes. If the crowns were simple shapes, such as cubes, then it would be easy to calculate their volume. But crowns are not simple shapes and it would have been almost impossible to measure them accurately enough for calculation purposes. This is where immersing them in water helps.

The crowns may have weighed the same in air but when Archimedes weighed the crowns immersed in water he observed that they had different weights. Putting this another way, each crown experienced a different loss in weight due to the buoyancy effect of the water. It is this *loss in weight* that was the key to solving the mystery. By measuring the loss in weight of the crowns, Archimedes was indirectly measuring their volumes.

To understand this, imagine a crown is immersed in a container full of water up to the over-flow pipe (Figure 2.16a). The crown displaces the water, spilling it down the over-flow where it is caught in another container. The volume of the spillage water can easily be measured and it has exactly the same volume as the crown. But the most interesting point is that the weight of the spillage water (water displaced) is equal to the loss in the weight of the crown. So, by measuring the loss in weight, Archimedes was in fact measuring the weight of displaced water, that is, the weight of an equal volume of water. As the weight density of water is a fixed value $(9,810 \text{ N/m}^3)$, it is a simple matter to convert this weight of water into a volume and so determine the density of the crown.

This is the principle that Archimedes discovered: When an object is immersed in water it experiences a loss in weight and this is equal to the weight of water it displaces.

What Archimedes measured was not actually the density of gold but its relative density or specific gravity as it is more commonly known. This is the density of gold relative to that of water and he calculated this using the formula

$$\text{specific gravity} = \frac{\text{weight of crown}}{\text{weight loss when immersed in water}}.$$

This may not look like the formula for specific gravity in Section 1.11.2 but it is the same. From Section 1.11.2

$$\text{specific gravity} = \frac{\text{weight of an object}}{\text{weight of an equal volume of water}}.$$

But Archimedes principle states that

weight loss when immersed in water = weight of an equal volume of water.

So, the two formulae are in fact identical and Archimedes was able to tell whether the crown was made of gold or not by some ingenious thinking and some simple calculations. The method works for all materials and not just gold and also for all fluids and not just water. Indeed, this immersion technique is now a standard laboratory method for measuring the volume of irregular shaped objects and for determining their specific gravity.

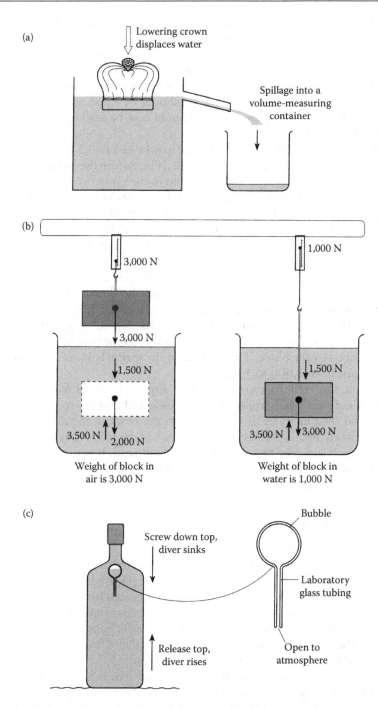

Figure 2.16 Archimedes principle. (a) Measuring the volume of an irregular object. (b) Demonstrating apparent loss in weight. (c) The 'Cartesian Diver'.

Still not convinced? Try this example with numbers. A block of material has a volume of 0.2 m³ and is suspended on a spring balance (Figure 2.16b) and weighs 3,000 N. When the block is lowered into the water it displaces 0.2 m³ of water. As the water weighs 10,000 N/m³ (approximately), the displaced water weighs 2,000 N (i.e. 0.2 m³ × 10,000 N/m³). Now, according to Archimedes, the weight of this water should be equal to the weight loss by the block and so the spring balance should now be reading only 1,000 N (i.e. 3,000–2,000 N).

To explain this, think about the space that the block (0.2 m³) will occupy when it is lowered into the water (Figure 2.16b). Before the block is lowered into the water, the 'space' it would take up is currently occupied by 0.2 m³ of water weighing 2,000 N. Suppose that the water directly above the block weighs 1,500 N (note that any number will do for this argument). Adding the two weights together is 3,500 N and this is supported by the underlying water and so there is an equal and upward balancing force of 3,500 N. The block is now lowered into the water and it displaces 0.2 m³ of water. The water under the block takes no account of this change and continues to push upwards with a force of 3,500 N and the downward force of the water above it continues to exert a downward force of 1,500 N. The block thus experiences a net upward force or a loss in weight of 2,000 N (i.e. 3,500–1,500 N). This is exactly the same value as the weight of water that was displaced by the block. The reading on the spring balance is reduced by this amount from 3,000 to 1,000 N.

A simple but striking example of this apparent weight loss is to tie a length of cotton thread around a house brick and first try to suspend it in the air and then in water. If you try to lift the brick in the air, the thread will very likely break. But the uplift force when the brick is in water means that the brick can now be lifted easily by the thread. It is this same weight 'loss' that enables rivers to move great boulders during floods and the sea to move shingle along the beach.

2.12.1 Floating objects

When an object such as a cork floats on water it is supported by the uplift force or buoyancy. It appears that the object has *lost* all of its weight. If the cork was held below the water surface and then released, it rises to the surface. This is because the weight of the water displaced by the cork is greater than the weight of the cork itself and so the cork rises under the unbalanced force. Once at the surface, the weight of the cork is balanced by the lifting effect of the water. In this case, the water displaced by the cork *is not a measure of its volume but a measure of its weight.*

Another way of determining if an object will float is to measure its density. When the density is less than that of water it will float. When it is greater it will sink. A block of wood, for example, is half the density of water and so it floats half submerged. Icebergs, which have a density close to that of water, float with only one tenth of their mass above the surface. The same principle also applies to other fluids. Hydrogen balloons, for example, rise in air because hydrogen is 14 times less dense than air.

Steel is six times denser than water and so it will sink. People laughed when it was first proposed that ships could be made of steel and would float. But today we just take such things for granted. Ships float even when loaded because much of their volume is filled with relatively light cargo and a lot of air space and so their average density is less than that of sea water.

Buoyancy is also affected by the density of seawater, which varies considerably around the world and affects the load that ships can safely carry. In Bombay, the sea is more salty than it is near the United Kingdom and so ships ride higher in the water. If a ship is loaded with cargo in Bombay and is bound for London, as it nears the United Kingdom it will lie much lower in the water and this could be dangerous if it is overloaded.

The 'Cartesian diver' is an interesting example of an object, which can either sink or float by slightly varying its density a little above or below that of water (Figure 2.16c). The 'diver' is really a small length of glass tubing, sealed and blown into a bubble at one end and open at the other. You can easily make one in a laboratory using a bunsen burner and a short length of glass tube. Next, find a bottle with a screw top, fill it with water and put the diver into the water. The diver will float because the air bubble ensures that the average density is less than that of water. Now, screw down the top and the diver will sink. This is because this action increases the water pressure, which compresses the air in the diver and increases its average density above that of water. Releasing the screw top allows the diver to rise to the surface again. This same principle is used to control submarines. When a submarine dives, its tanks are allowed to fill with water so that its average density is greater than that of water. The depth of submergence is determined by the extent to which its tanks are flooded. To make the submarine rise, water is blown out of its tanks using compressed air.

To summarise:

An object floats when it is less dense than water but sinks when it is denser than water. When an object floats it displaces water equal to its weight.

2.12.2 Submarine problem

Here is a problem to test your understanding of Archimedes principle.

A submarine is floating in a lock (Figure 2.17). It then submerges and sinks to the bottom. What happens to the water level in the lock? Does it rise or fall?

Archimedes principle says that when an object floats it displaces its own weight of water and when it sinks it displaces its own volume.

Applying this to the submarine – when it is floating on the water surface, the submarine displaces its own weight of water which will

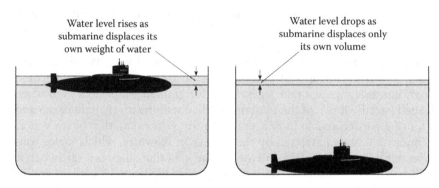

Water level rises as
submarine displaces its
own weight of water

Water level drops as
submarine displaces only
its own volume

Figure 2.17 Applying Archimedes principle to submarines.

be substantial because submarines are heavy. But when the submarine sinks to the bottom it only displaces water equivalent to its volume. So, the amount of water displaced by the floating submarine will be much greater than the volume of water displaced when it is submerged. This means, that when the submarine dives, the water level in the lock will drop (very slightly!).

2.12.3 Ice problem

Another interesting problem occurs when ice is added to water. When ice is added, the water level rises. But when the ice melts what happens to the water level? Does it rise, fall or stay the same?

Ice is a solid object that floats and so it should behave in the same way as any other solid object. When it melts, however, it becomes part of the water and, in effect, it 'sinks'.

To see what happens, take 1.0 L of water, which has a mass of 1 kg and weight 10 N, and freeze it. Water expands as it freezes and so as it turns into ice its volume will increase by approximately 8%; to 1.08 L of ice. But remember it is still only 1.0 L of water and so its weight has not changed – just its volume. If the block of ice is now put into the tank it will float on the water, like an iceberg, because the density of the ice is slightly less than the water. The water level in the tank will also rise as a result of the displacement by the ice – like any solid object that floats, the ice displaces its own weight of water, which is still 10 N. Now, 10 N of water has a volume of 1.0 L and so 1.0 L of water will be displaced. It has nothing to do with the volume of the ice – only its weight. However, when the ice melts it 'sinks' into the tank, and like any other object that sinks, it now displaces its own volume of water. But the volume of the melted ice (now water again) is 1.0 L. So, the displacement in each case is the same – 1.0 L – which means the water level in the tank remains unchanged when the ice melts. Try this for yourself with a jug of water and some ice cubes (Figure 2.18).

Figure 2.18 Ice problem.

It is the volume of the ice that can mislead your thinking because it changes significantly when the water freezes. Do not be misled by this, just follow the principle of Archimedes and everything will work out right.

This principle can be applied to the concerns over melting ice caps and sea level rises as we begin to experience changes in our climate. If the North Polar ice melts, the sea level would not be affected because the ice is already in the sea and is floating – just like the above example. But if the South Polar ice melts, this would cause the sea level to rise because the ice is on land and would add to the volume of water in the oceans.

2.12.4 Drowning in quicksand – myth or reality?

A common scene in many adventure films is of someone stumbling into a patch of quicksand and getting sucked under (Figure 2.19). Great drama but is this what really happens? Recent research at the Ecole Normal Supérieure in Paris, based partly on Archimedes principle, suggests otherwise. Apparently, the work was inspired by a holiday trip to the legendary quicksand at Daryacheh-ye Namak salt lake near Qom in Iran where local shepherds speak of camels disappearing without trace.

Quicksand is a mixture of fine sand, clay and salt water, in which the grains are delicately balanced and very unstable. This makes the mixture appear solid but once it is disturbed it starts to behave like a liquid and so if you stand on it you will start to sink into it. But just like any liquid, there is a

Figure 2.19 Quicksand problem.

buoyancy effect (Archimedes principle) and so how far you will sink depends on your density. So, if you float in water you will also float in quicksand.

But the problem is not just a hydraulic one. The research showed that when the mixture liquefies, the sand and clay fall to the bottom and create thick sediment that also helps to prevent you sinking further. So, the good news is that the two combined mean that you are unlikely to sink much beyond your waist – though not such good news if you fell in head first. Struggling and kicking will not make you sink further – it just makes the mixture more unstable and so you will sink faster. The bad news comes when you try to get out because the mixture will hold you fast. It can take as much force to pull you out of quicksand as it does to lift a typical family car. So, you are more likely to have your limbs pulled off than get out of the mess! So, how do you get out? One suggestion by the researchers is to gently wriggle your feet to liquefy the mixture and then slowly pull yourself up a few millimetres at a time.

The myth surrounding quicksand probably originates from people falling in head first and in such circumstances you are most likely to drown.

Science also spoils a good story – it means that all those film dramas about quicksand such as *The Hound of the Baskervilles* are pure fantasy!

2.13 SOME EXAMPLES TO TEST YOUR UNDERSTANDING

1. Determine the pressure in kN/m² for a head of (a) 14 m of water and (b) 1.7 m of oil. Assume the mass density of water is 1,000 kg/m³ and oil is 785 kg/m³ (137.34 kN/m², 13.09 kN/m²).
2. A storage tank, 2.3 m long by 1.2 m wide and 0.8 m deep is full of water. Calculate (a) the mass of water in the tank, (b) the pressure on the bottom of the tank, (c) the force on the end of the tank and (d) the position of this force below the water surface (2,210 kg, 7,848 N/m², 3,767 N, 0.53 m below the water surface).
3. Calculate atmospheric pressure in kN/m² when the barometer reading is 750 mm of mercury. Calculate the height of a water barometer needed to measure atmospheric pressure (100.06 kN/m², 10.2 m).
4. Calculate the pressure in kN/m²and in m head of water in a pipeline carrying water using a mercury manometer when $h_1 = 0.5$ m and $h_2 = 1.2$ m. Assume the specific gravity of mercury is 13.6 (155 kN/m², 15.82 m).
5. A vertical rectangular sluice gate 1.0 m high by 0.5 m wide is used to control the discharge from a storage reservoir. Calculate the horizontal force on the gate and its location in relation to the water surface when the top of the gate is located 2.3 m below the water surface (13.73 kN, 2.83 m).
6. Calculate the force and its location below the water surface on a 0.75 m diameter circular sluice gate located when the top of the gate is located 2.3 m below the water surface (11.57 kN, 2.69 m).

Chapter 3

When water starts to flow

Hydrodynamics

3.1 INTRODUCTION

Hydrodynamics is the study of water flow. It helps us to understand how water behaves when it flows in pipes and channels and to answer such questions as – What diameter of pipe is needed to supply a village or a town with water? How wide and deep must a channel be to carry water from a dam to an irrigation scheme? What kind of pumps might be required and how big they must be? These are the practical problems of hydrodynamics.

Hydrodynamics is more complex than hydrostatics because it must take account of more factors, particularly the direction and velocity in which the water is flowing and the influence of viscosity. In early times hydrodynamics, like many other developments, moved forward on a trial-and-error basis. If the flow was not enough, then a larger diameter pipe was used; if a pipe bursts under the water pressure, then a stronger one was put in its place. But during the past 300 years or so, scientists have found new ways of answering the questions about size, shape and strength. They experimented in laboratories and came up with mathematical theories that have now replaced trial-and-error methods for solving the most common hydrodynamic problems.

3.2 EXPERIMENTATION AND THEORY

Experimentation was a logical next step from trial and error. Scientists built physical models of hydraulic systems in the laboratory and tested them before building the real thing. Much of our current knowledge of water flow in pipes and open channels has come from experimentation and devising empirical formulae which link water flow with the size of pipes and channels. Today, we use formulae for most design problems, but there are still some problems which are not easily solved in this way. Practical laboratory experiments are still used to find solutions for the design of complex works such as harbours, tidal power stations, river flood control schemes and dam spillways. Small-scale models are built to test new designs; though mathematical modelling is beginning to take over as a cheaper option (Figure 3.1).

Some formulae have also been developed analytically from our understanding of the basic principles of physics, which include the properties of water and Newton's laws of motion. The design rules for hydrostatics were developed analytically and have proved to work very well. But when water starts to move, it is difficult to take account of all the new factors involved, in particular viscosity. The engineering approach, rather than the scientific one, is to try and simplify a problem by ignoring those forces which do not seriously impact the outcome. Viscosity is usually ignored because its effects are small. This greatly simplifies problems and solutions. Ignoring the forces of viscosity makes pipeline design much simpler. It has no effects on the final choice of pipe size. Other more important factors dominate the design process such as velocity, pressure and the forces of friction. These do have significant influence on the choice of pipe size and so it is important to focus attention on them. This is why engineering is often regarded as much an art as a science. The science is about knowing what physical factors must be taken into account but the art of engineering is knowing which of the factors can be safely ignored in order to simplify a problem without it seriously affecting the accuracy of the outcome.

Remember, engineers are not always looking for high levels of accuracy. There are inherent errors in all the data and so there is little point in calculating the diameter of a pipe to several decimal places when the data being

Figure 3.1 Laboratory model of a dam spillway.

used have not been recorded with the same precision. Electronic calculators and computers have created much of this problem and many students still continue to quote answers to many decimal places simply because the computer says so. The answer is only as good as the data going into the calculation and so another engineering skill is to know how accurate an answer needs to be. Unfortunately, many of the arts of engineering can only be learned through practice and experience. This is the reason why a vital part of training young engineers always involves working with older, more experienced engineers to acquire the skills. Just knowing the right formula is not enough.

The practical issues of cost and availability also impose limitations on hydraulic designs. For example, commercially available pipes come in a limited range of sizes, for example, 50, 75, 100 mm diameter. If an engineer calculates that a 78 mm diameter pipe is needed, he is likely to choose the next size of pipe to make sure it will do the job properly, that is, 100 mm. So, there is nothing to be gained in spending a lot of time refining the design process in such circumstances.

Simplifying problems so that they can be solved more easily, without loss of accuracy, is at the heart of hydrodynamics – the study of water movement.

3.3 HYDRAULIC TOOLBOX

The development of hydraulic theory has produced *three* important basic tools (equations) which are fundamental to solving most hydrodynamic problems:

- Discharge and continuity
- Energy
- Momentum

They are not difficult to master and you will need to understand them well.

3.4 DISCHARGE AND CONTINUITY

Discharge refers to the volume of water flowing along a pipe or channel each second. Volume is measured in cubic metres (m^3) and discharge is measured in cubic metres per second (m^3/s). Alternative units are litres per second (L/s) and cubic metres per hour (m^3/h).

There are two ways of determining discharge. The first involves measuring the volume of water flowing in a system over a given time period. For example, water flowing from a pipe can be caught in a bucket of known volume (Figure 3.2a). If the time to fill the bucket is recorded, the discharge from the pipe can be determined using the following formula

$$\text{discharge (m}^3\text{/s)} = \frac{\text{volume (m}^3)}{\text{time (s)}}.$$

Discharge can also be determined by multiplying the velocity of the water by the area of the flow. To understand this, imagine water flowing along

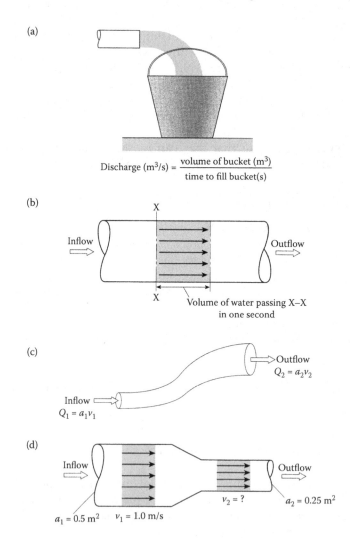

(a)

$$\text{Discharge (m}^3\text{/s)} = \frac{\text{volume of bucket (m}^3)}{\text{time to fill bucket(s)}}$$

(b)

Inflow

Outflow

X

X Volume of water passing X–X in one second

(c)

Outflow $Q_2 = a_2 v_2$

Inflow $Q_1 = a_1 v_1$

(d)

Inflow

Outflow

$v_2 = ?$

$a_2 = 0.25 \text{ m}^2$

$a_1 = 0.5 \text{ m}^2$ $v_1 = 1.0 \text{ m/s}$

Figure 3.2 Discharge and continuity. (a) Measuring discharge using volume and time. (b) Measuring velocity and area of flow. (c) Continuity inflow equals outflow. (d) Example in Box 3.1.

a pipeline (Figure 3.2b). In 1 s, the volume of water flowing past x–x will be the shaded volume. This volume can be calculated by multiplying the area of the pipe by the length of the shaded portion. But the shaded length is numerically equal to the velocity v and so the volume flowing each second (i.e. the discharge) is equal to the pipe area multiplied by the velocity. Writing this as an equation

discharge (Q) = velocity (v) × area (a)

$$Q = va.$$

The *continuity equation* builds on the discharge equation and simply means that the amount of water flowing into a system must be equal to the amount of water flowing out of it (Figure 3.2c).

inflow = outflow.

And so

$$Q_1 = Q_2.$$

But from the discharge equation

$$Q = va$$

Substituting in the continuity equation for Q

$$v_1 \, a_1 = v_2 \, a_2.$$

So the continuity equation not only links discharges but also links areas and velocities as well. This is a very simple but powerful equation and is fundamental to solving many hydraulic problems. An example in Box 3.1 shows how this works in practice for a pipeline with a changing diameter.

The simple equation of inflow equals outflow is only true when the flow is steady. This means the flow remains the same over time. But there are situations when the inflow does not equal the outflow. An example of this occurs in our homes (Figure 3.3). The tank in the attic stores water so that the inflow into our homes does not limit our water demand at peak times. On a much larger scale, reservoirs on rivers perform a similar function for water supply schemes – balancing supply and demand at peak times (see Section 8.3.1). In these circumstances, a storage term is added to the continuity equation

inflow = outflow + rate of increase (or decrease) in storage.

Note that this equation is about discharges and so storage is shown as the rate at which it is increasing or decreasing. Hydrologists use a similar equation when calculating rainfall and run-off in the water basins. This is

---

BOX 3.1   EXAMPLE: CALCULATING VELOCITY USING THE CONTINUITY EQUATION

A pipeline changes area from 0.5 to 0.25 m^2 (Figure 3.2d). If the velocity in the larger pipe is 1.0 m/s, calculate the velocity in the smaller pipe.

Use the continuity equation

inflow = outflow.

And so

$v_1 a_1 = v_2 a_2$
$1 \times 0.5 = v_2 \times 0.25$
$v_2 = 2$ m/s.

Note how water moves much faster in the smaller pipe.

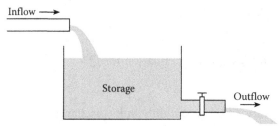

inflow = outflow + rate of increase (or decrease) in storage

Figure 3.3 Continuity when there is water storage.

the *water balance equation*. Water volumes are used rather than discharges (see Section 9.2.1).

3.5 ENERGY

The second basic tool is the energy equation. This links pressure and velocity. Energy is the capacity to do useful work (see Section 1.9). Water can have three kinds of energy:

• Pressure energy
• Kinetic energy
• Potential energy

Energy for solid objects has the dimensions of Nm. For fluids, the units are little different. It is common practice to measure energy in terms of *energy per unit weight* and so the unit of water energy is Nm/N. The Newton terms cancel each other out and we are left with metres (m). This makes energy look similar to pressure head; both are measured in metres. Indeed, we shall see that the terms energy and pressure head are in fact interchangeable. So, let's explore these three types of water energy.

3.5.1 Pressure energy

When the water is under pressure, it can do useful work for us. Water released from a tank can drive a small turbine which in turn drives a generator to produce electrical energy (Figure 3.4a). So, the pressure available in the tank is a measure of the energy available to do work. It is calculated as follows

$$\text{pressure energy} = \frac{p}{\rho g},$$

where p is pressure (kN/m^2), ρ is mass density (kg/m^3) and g is the gravity constant (9.81 m/s^2).

Notice that the equation for pressure energy is actually the same as the familiar pressure – head equation (remember $p = \rho g h$). It is just presented in a different way. So pressure energy is in fact the same as the pressure head and is measured in metres (m).

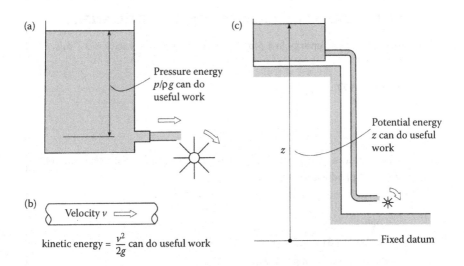

Figure 3.4 Measuring energy. (a) Pressure energy. (b) Kinetic energy. (c) Potential energy.

3.5.2 Kinetic energy

When the water flows, it possesses energy because of this movement and this is known as *kinetic energy* – sometimes called *velocity energy*. The faster the water flows the greater is its kinetic energy (Figure 3.4b). It is calculated as follows

$$\text{kinetic energy} = \frac{v^2}{2g}$$

where v is velocity (m/s) and g is gravity constant (9.81 m/s²).

Kinetic energy is also measured in metres (m) and for this reason it is usually called the *velocity head*. An example of how to calculate kinetic energy is shown in Box 3.2.

3.5.3 Potential energy

Water also has energy as a result of its location. Water stored in the mountains can do useful work by generating hydropower, whereas water stored on a flood plain has little or no potential for work (Figure 3.4c). The higher the water source the more energy water has. This is called *potential energy*. It is determined by the height of the water in metres above some fixed datum point

$$\text{potential energy} = z,$$

BOX 3.2   EXAMPLE: CALCULATING KINETIC ENERGY

Calculate the kinetic energy in a pipeline when the flow velocity is 3.7 m/s.

$$\text{kinetic energy} = \frac{v^2}{2g}$$

$$= \frac{3.7^2}{2 \times 9.81} = 0.7 \text{ m}$$

Think of this as a velocity head so calculate the equivalent pressure in kN/m² that would produce this kinetic energy.

To calculate velocity head as a pressure in kN/m² use

$$p = \rho g h$$

$$= 1,000 \times 9.81 \times 0.7$$

$$= 6,867 \text{ N/m}^2 = 6.87 \text{ kN/m}^2.$$

---

where z is the height of the water in metres (m) above a fixed datum. Measuring potential energy must be relative to some fixed point to have meaning. It is similar to using sea level as the fixed datum for measuring changes in land elevation.

3.5.4 Total energy

The really interesting point of all this is that not only are all the different forms of energy interchangeable (pressure energy can be changed to velocity energy as so on) but they can also be added together to help us solve a whole range of hydrodynamic problems. The Swiss mathematician Daniel Bernoulli (1700–1782) made this most important discovery. Indeed, it was Bernoulli who is said to have put forward the name of hydrodynamics to describe water flow. This led to one of the best known equations in hydraulics – *total energy equation*. It is often called the *Bernoulli equation* in recognition of his contribution to the study of fluid behaviour.

The total energy in a system is the sum of all the different energies

$$\text{total energy} = \frac{p}{\rho g} + \frac{v^2}{2g} + z$$

On its own, simply knowing the total energy in a system is of limited value. But realising that the total energy will be the same throughout a system, even though the various components of energy may be different, makes it much more useful. Take, for example, water flowing in a pipe from point 1 to point 2 (Figure 3.5). The total energy at point 1 will be the same as the total energy at point 2. So, we can rewrite the total energy equation in a different and more useful way

total energy at point 1 = total energy at point 2.

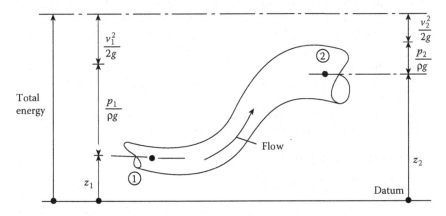

Figure 3.5 Total energy is the same throughout the system.

Writing the equation in mathematical terms

$$\frac{p_1}{\rho g} + \frac{v_1^2}{2g} + z_1 = \frac{p_2}{\rho g} + \frac{v_2^2}{2g} + z_2$$

The velocity, pressure and height at 1 are all different from those at point 2 but when they are added together at each point the total is the same. This means that if we know some of the values at point 1, we can now predict values at point 2. There are examples of this in the next section.

Note that the energy equation only works for the flows where there is little or no energy loss between the points being considered. However, it is a reasonable assumption to make in many situations, though not so reasonable for long pipelines where energy losses can be significant and so cannot be ignored. But for now, assume that water is an ideal fluid and that no energy is lost. Later, in Chapters 4 and 5, we will see how energy losses can be incorporated into the equation.

3.6 SOME USEFUL APPLICATIONS OF THE ENERGY EQUATION

The usefulness of the energy equation is well demonstrated in the following examples.

3.6.1 Pressure and elevation changes

The total energy equation tells us that pressures in pipelines change with elevation. Pipelines tend to follow the natural ground contours up and down the hills. As a result, pressure changes simply as a result of differences in ground levels. So, a pipeline designer must be fully aware of the terrain over which the pipeline runs in order to deliver the right pressure. For example, a pipeline running uphill will experience a drop in pressure of 10 m head for every 10 m rise in ground level. Similarly, the pressure in a pipe running downhill will increase by 10 m for every 10 m fall in ground level. The energy equation explains why this is so.

A pipeline runs from a reservoir over undulating land. Consider total energy at two points 1 and 2 along the pipeline some distance apart and at different elevations (Figure 3.6).

Assuming no energy losses between these two points, the total energy in the pipeline at point 1 is equal to the total energy at point 2.

total energy at 1 = total energy at 2

$$\frac{p_1}{\rho g} + \frac{v_1^2}{2g} + z_1 = \frac{p_2}{\rho g} + \frac{v_2^2}{2g} + z_2$$

z_1 and z_2 are measured from some chosen horizontal datum.

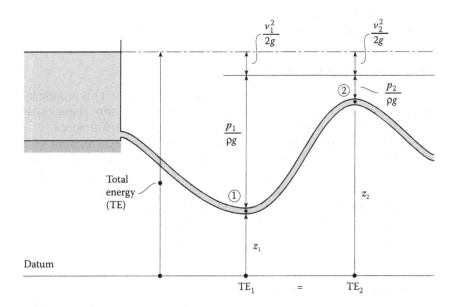

Figure 3.6 Pressure changes with elevation.

Normally, pipelines would have the same diameter and so the velocity at point 1 is the same as the velocity at point 2. This means that the kinetic energy at points 1 and 2 are also the same. The above equation then simplifies to

$$\frac{p_1}{\rho g} + z_1 = \frac{p_2}{\rho g} + z_2$$

Rearranging this to bring the pressure terms and the potential terms together

$$\frac{p_1}{\rho g} - \frac{p_2}{\rho g} = z_2 - z_1$$

Putting this into words

changes in pressure (m) = changes in ground level (m).

Here, p_1 and p_2 represent a pressure change between points 1 and 2 (measured in metres) which is a direct result of the change in ground level from z_1 to z_2. Note that this has nothing to do with the pressure loss due to friction as if often thought – it is just ground elevation changes. An example of how to calculate changes in the pressure due to changes in the ground elevation is shown in Box 3.3.

3.6.2 Measuring velocity

The energy equation is a useful tool for measuring velocity. This is done by stopping a small part of the flow and measuring the pressure change that results from this. Airline pilots use this principle to measure airspeed.

When the water (or air) flows around an object (Figure 3.7a), most of it is deflected around it but there is one small part of the flow which hits the object head-on and stops. Stopping the water in this way is called *stagnation* and the point at which this occurs is the *stagnation point*. Applying the energy equation to the main stream and the stagnation point

$$\frac{p_1}{\rho g} + \frac{v_1^2}{2g} + z_1 = \frac{p_s}{\rho g} + \frac{v_s^2}{2g} + z_s$$

Assuming the flow is horizontal

$$z_1 = z_s.$$

As the water stops at the stagnation point

$$v_s = 0,$$

and so

$$\frac{p_1}{\rho g} + \frac{v_1^2}{2g} = \frac{p_s}{\rho g}$$

Rearranging this equation to bring all the velocity and pressure terms together

$$\frac{v_1^2}{2g} = \frac{p_s}{\rho g} - \frac{p_1}{\rho g}$$

Rearranging it again for an equation for velocity v_1

$$v_1 = \sqrt{2\left(\frac{p_s - p_1}{\rho}\right)}$$

By measuring pressure in the mainstream p_1 and the pressure at the stagnation point p_s, it is possible to calculate the main stream velocity. This idea

BOX 3.3 EXAMPLE: CALCULATING PRESSURE CHANGES DUE TO ELEVATION CHANGES

A pipeline is constructed across undulating ground (Figure 3.6). Calculate the pressure at point 2 when the pressure at point 1 is 150 kN/m² and the elevation of point 2 is 7.5 m above point 1.

Assuming no energy loss along the pipeline, this problem can be solved using the energy equation

total energy at 1 = total energy at 2

$$\frac{p_1}{\rho g} + \frac{v_1^2}{2g} + z_1 = \frac{p_2}{\rho g} + \frac{v_2^2}{2g} + z_2.$$

As the pipe diameter is the same throughout, the velocity will also be the same, as will the kinetic energy. So the kinetic energy terms on each side of the equation cancel each other out.

The equation simplifies to

$$\frac{p_1}{\rho g} + z_1 = \frac{p_2}{\rho g} + z_2.$$

Rearranging the components to calculate the pressure at 2

$$\frac{p_1}{\rho g} - \frac{p_2}{\rho g} = z_2 - z_1$$

All elevation measurements are made from the same datum level and so

$$z_2 - z_1 = 7.5 \text{ m}.$$

This means that

$$\frac{p_1 - p_2}{\rho g} = 7.5 \text{ m}$$

and so

$$p_1 - p_2 = 1{,}000 \times 9.81 \times 7.5$$
$$= 73{,}575 \text{ N/m}^2 = 73.6 \text{ kN/m}^2$$

known pressure at point 1 = 150 kN/m²

and so

pressure at point 2 = 150 − 73.6 = 76.4 kN/m².

So, the energy equation predicts a drop in pressure at point 2 which is directly attributed to the elevation rise in the pipeline.

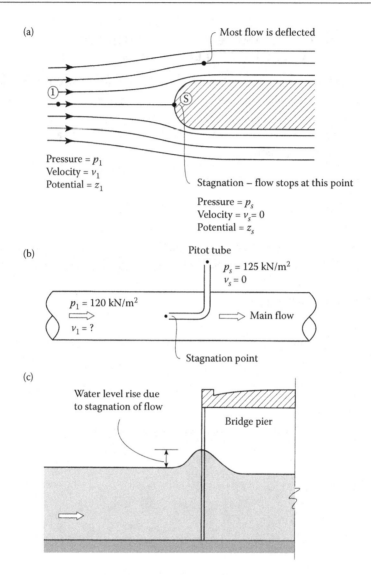

(a)

Most flow is deflected

①

Pressure = p_1
Velocity = v_1
Potential = z_1

Stagnation – flow stops at this point

Pressure = p_s
Velocity = $v_s = 0$
Potential = z_s

(b)

Pitot tube

$p_s = 125$ kN/m^2
$v_s = 0$

$p_1 = 120$ kN/m^2

$v_1 = ?$

Main flow

Stagnation point

(c)

Water level rise due
to stagnation of flow

Bridge pier

Figure 3.7 Measuring velocity using stagnation points. (a) Stagnation point when water flows around an object. (b) Measuring velocity with a pitot tube. (c) Stagnation point in front of a bridge pier.

is used widely for measuring water velocity in the pipes using a device known as a *pitot tube* (Figure 3.7b). The stagnation pressure p_s on the end of the tube is measured together with the general pressure in the pipe p_1. The velocity is then calculated using the energy equation (see Box 3.4). One disadvantage of this device is that it does not measure the average velocity in a pipe but only the velocity at the point where the pitot tube is located.

However, this can be very useful for experimental work that explores the changes in velocity across the diameter of a pipe to produce velocity profiles. Pitot tubes are also used on aircraft to measure the aircraft's velocity. As both the air and the aircraft are moving, the pilot will adjust the velocity reading to take account of this.

BOX 3.4   EXAMPLE: CALCULATING THE VELOCITY IN A PIPE USING A PITOT TUBE

Calculate the velocity in a pipe using a pitot tube when the normal pipe operating pressure is 120 kN/m^2 and the pitot pressure is 125 kN/m^2 (Figure 3.7b).

Although there is an equation for velocity given in this text, it is a good idea at first to work from basic principles to build up your confidence in its use. The problem is solved using the energy equation. Point 1 describes the main flow and point s describes the stagnation point on the end of the pitot tube

$$\frac{p_1}{\rho g} + \frac{v_1^2}{2g} + z_1 = \frac{p_s}{\rho g} + \frac{v_s^2}{2g} + z_s$$

At the stagnation point

$$v_s = 0.$$

And as the system is horizontal

$$z_1 = z_s = 0.$$

This reduces the energy equation to

$$\frac{p_1}{\rho g} + \frac{v_1^2}{2g} = \frac{p_s}{\rho g}$$

All the values in the equation are known except for v_1 so calculate v_1

$$\frac{120,000}{1,000 \times 9.81} + \frac{v_1^2}{2 \times 9.81} = \frac{125,000}{1,000 \times 9.81}$$

$$12.23 + \frac{v_1^2}{2g} = 12.74$$

$$\frac{v_1^2}{2 \times 9.81} = 12.74 - 12.23 = 0.51\,m$$

$$v_1 = \sqrt{2 \times 9.81 \times 0.51} = 3.16\,m/s$$

Stagnation points also occur in the channels. One example occurs at a bridge pier (Figure 3.7c). Notice how the water level rises a little just in front of the pier as the kinetic energy in the river changes to pressure energy as the flow stops. In this case, the pressure rise is seen as a rise in water level. Although this change in the water level could be used to calculate the velocity of the river, it is rather small and difficult to measure accurately. So, it is not a very reliable way of measuring velocity in channels.

3.6.3 Orifices

Orifices are usually gated openings at the bottom of tanks and reservoirs used to control the release of water flow into a channel or some other collecting basin (Figure 3.8a). They are mostly rectangular or circular openings. The energy equation makes it possible to calculate the discharge released through an orifice by first calculating the flow velocity from the orifice and then multiplying it by the area of the opening. One important proviso at this stage is that the orifice must discharge freely

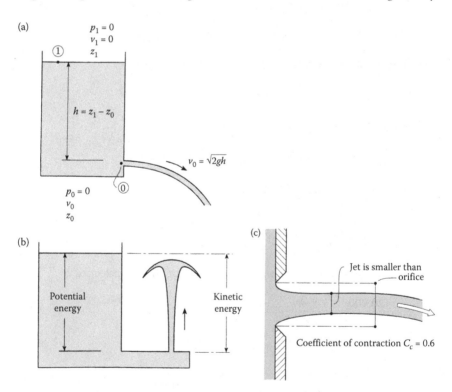

$$p_1 = 0$$
$$v_1 = 0$$
$$z_1$$

$$h = z_1 - z_0$$

$$v_0 = \sqrt{2gh}$$

$$p_0 = 0$$
$$v_0$$
$$z_0$$

Potential energy

Kinetic energy

Jet is smaller than orifice

Coefficient of contraction $C_c = 0.6$

Figure 3.8 Flow through orifices. (a) Typical orifice flow. (b) Changing potential energy into kinetic energy. (c) Flow contraction as it leaves an orifice.

and unhindered into the atmosphere, otherwise this approach will not work. Some orifices do operate in submerged conditions and this does affect the flow.

The energy equation for an orifice in a tank (Figure 3.8a) is

$$\frac{p_1}{\rho g} + \frac{v_1^2}{2g} + z_1 = \frac{p_0}{\rho g} + \frac{v_0^2}{2g} + z_0$$

Note the careful choice of the points for writing the energy terms. Point 1 is chosen at the water surface in the tank and point 0 is at the centre of the orifice.

At the water surface, the pressure is atmospheric and so is assumed to be zero (remember all pressures are measured relative to atmospheric pressure which is taken as the zero point). As the water in the tank is not moving, the kinetic energy is also zero. So, all the initial energy is potential. At the orifice, the jet comes out into the atmosphere and as the jet does not burst open it is assumed that the pressure in and around the jet is atmospheric pressure, that is, zero. So, the equation reduces to

$$z_1 = \frac{v_0^2}{2g} + z_0$$

Rearranging this equation

$$\frac{v_0^2}{2g} = z_1 - z_0 = h,$$

where h is the depth of water from the surface to the centre of the orifice. Now, rearrange the equation to calculate v_0

$$v_0 = \sqrt{2gh}.$$

Evangelista Torricelli (1608–1647) first made this connection between the pressure head available in the tank and the velocity of the emerging jet some considerable time before Bernoulli developed his energy equation. As a pupil of Galileo, he was greatly influenced by him and applied his concepts of mechanics to water falling under the influence of gravity. Although, the above equation is now referred to as Torricelli's law, he did not include the $2g$ term. This was introduced much later by other investigators.

Torricelli sought to verify this law by directing a water jet from an orifice, vertically upwards (Figure 3.8b). He showed that the jet rose to almost the same height as the free water surface in the tank showing that the potential energy in the tank and the velocity energy at the orifice were equal. So knowing the pressure head available in a pipe, it is possible to calculate the height to which a water jet would rise if a nozzle was attached to it – very useful for designing fountains!

The velocity of a jet can also be used to calculate the jet discharge using the discharge equation

$$Q = av.$$

So

$$Q = a\sqrt{2gh}.$$

The area of the orifice a is used in the equation because it is easy to measure but this means the end result is not so accurate because the area of the jet of water is not the same as the area of the orifice. As the jet emerges and flows around the edge of the orifice, it follows a curved path and so the jet ends up smaller in diameter than the orifice (Figure 3.8c). The contraction of the jet is taken into account by introducing a *coefficient of contraction* C_c which has a value of approximately 0.6. So, the discharge formula now becomes

$$Q = C_c a\sqrt{2gh}.$$

Although it might be interesting to work out the discharge from holes in tanks, a more useful application of Torricellis's law is the design of underflow gates for both measuring and controlling discharges in open channels (see Section 6.2).

3.6.4 Pressure and velocity changes in a pipe

A more general and very practical application of the energy equation is to predict pressures and velocities in pipelines as a result of changes in ground elevation and pipe sizes. An example in Box 3.5 shows just how versatile this equation can be.

3.7 SOME MORE ENERGY APPLICATIONS

3.7.1 Flow through narrow openings

When water flows through narrow openings in pipes and channels, such as valves or gates, there is a tendency to assume they are constricting the flow.

BOX 3.5 EXAMPLE: CALCULATING PRESSURE CHANGES USING THE ENERGY EQUATION

A pipeline carrying a discharge of 0.12 m³/s changes from 150 mm diameter to 300 mm diameter and rises through 7 m (Figure 3.9). Calculate the pressure in the 300 mm pipe when the pressure in the 150 mm pipe is 350 kN/m².

This problem involves changes in pressure, kinetic and potential energy, and its solution requires both the energy and continuity equations. The first step is to write down the energy equation for the two points in the systems 1 and 2

$$\frac{p_1}{\rho g} + \frac{v_1^2}{2g} + z_1 = \frac{p_2}{\rho g} + \frac{v_2^2}{2g} + z_2$$

The next step is to identify the known and unknown values in the equation. p_1, z_1 and z_2 are known ($z_2 - z_1 = 7$ m) values but p_2 is unknown and so are v_1 and v_2. First, determine v_1 and v_2 use the continuity equation

$$Q = va.$$

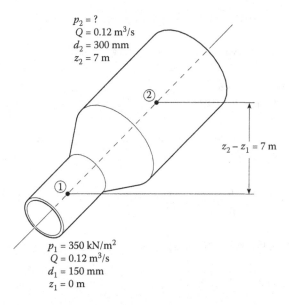

$p_2 = ?$
$Q = 0.12 \text{ m}^3/\text{s}$
$d_2 = 300 \text{ mm}$
$z_2 = 7 \text{ m}$

$z_2 - z_1 = 7 \text{ m}$

$p_1 = 350 \text{ kN/m}^2$
$Q = 0.12 \text{ m}^3/\text{s}$
$d_1 = 150 \text{ mm}$
$z_1 = 0 \text{ m}$

Figure 3.9 Calculating changes in pressure in a pipeline.

Rearranging this to calculate v

$$v = \frac{Q}{a}.$$

And so

$$v_1 = \frac{Q}{a_1} = \text{and} \quad v_2 = \frac{Q}{a_2}.$$

The pipe areas are not known but their diameters are known, so next calculate their cross sectional areas

$$a_1 = \frac{\pi d_1^2}{4} = \frac{\pi 0.15^2}{4} = 0.018 \, m^2$$

$$a_2 = \frac{\pi d_2^2}{4} = \frac{\pi 0.3^2}{4} = 0.07 \, m^2.$$

Now, calculate the velocities

$$v_1 = \frac{Q}{a_1} = \frac{0.120}{0.018} = 6.67 \, m/s$$

$$v_2 = \frac{Q}{a_2} = \frac{0.120}{0.07} = 1.71 \, m/s.$$

Putting all the known values into the energy equation

$$\frac{350,000}{1,000 \times 9.81} + \frac{6.67^2}{2 \times 9.81} + 0 = \frac{p_2}{\rho g} + \frac{1.71^2}{2 \times 9.81} + 7$$

Note the pressures in the equation are in N/m^2 and not in kN/m^2. The equation simplifies to

$$35.68 + 2.26 = \frac{p_2}{\rho g} + 0.15 + 7.$$

Rearranging this equation for p_2

$$\frac{p_2}{\rho g} = 35.68 + 2.26 - 0.15 - 7$$

$$= 30.8 \, m \text{ head of water.}$$

To determine this head as a pressure in kN/m² use the pressure-head equation

$$\text{pressure} = \rho g h$$
$$p_2 = 1{,}000 \times 9.81 \times 30.8$$
$$= 302{,}000 \text{ N/m}^2 = 302 \text{ kN/m}^2$$

But this is not always the case. The reason for this misunderstanding is that we live in a solid world and so we logically apply what we see to water. People cause chaos when too many try to get through a narrow opening at the same time. So surely water must behave in a similar way. Well this is where water surprises everyone – it behaves quite differently (Figure 3.10).

Continuity and energy control what happens when water flows along a pipe and meets a constriction. As the pipe becomes narrower, the water, rather than slowing down, actually speeds up. The continuity equation tells us that when the area is smaller the velocity must be greater. But surely the constriction must slow the whole discharge and hence the velocity. Well no – the discharge is governed by the total energy available to drive the flow and as there is no change in the total energy between the main pipe and the constriction, the discharge in the system does not change. So, the

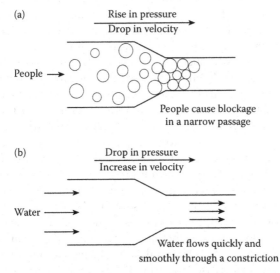

Figure 3.10 People and water flow differently through narrow passages. (a) People flow. (b) Water flow.

flow passes quickly and smoothly through the constriction without fuss. It would seem that water behaves much more sensibly than people!

What does happen of course is that the pressure in the system changes at the constriction. The kinetic energy increases and this results in a corresponding decrease in the pressure energy. So a narrow pipe, or indeed any other constriction such as a partly open valve, does not throttle the flow, it just speeds it up so that it goes through much faster. You can see this when you open and close a tap at home. The discharge through a partially open tap is almost the same as that through a fully open one. The total energy available is the same but the flow area is smaller when it is partially opened and so the water just flows through with a greater velocity. Of course, the velocity is eventually slowed when the tap is almost shut and at this point energy losses at the tap dominate the flow.

This same principle also applies to flow in open channels. When flow is constricted, it speeds up (kinetic energy increases) and the water level drops (pressure energy decreases). You can see this effect as water flows through channel constrictions at bridges and weirs (this is discussed in detail in Chapters 4 and 5).

Some people have suggested that the design of sports stadiums, which can easily become congested with people, could benefit from linking the flow of people to the flow of water. Some years ago, there was a major accident at a football stadium in Belgium in which many people were crushed to death when those at the rear of the stadium suddenly surged forward in a narrow tunnel pushing those in front onto fixed barriers and crushing them. At the time it was suggested that stadiums should, in future, be designed with hydraulics in mind so the layout, size and shape of tunnels and barriers would allow people to 'flow' smoothly onto the terraces in a more orderly and safe manner. This is a dangerous analogy because people do not 'flow' like water. They tend to get stuck in narrow passages and against solid barriers whereas water behaves much more sensibly, flowing around barriers and speeding up and slowing down when needed to get through the tight spots.

Research at Buckinghamshire New University undertaken to make sure the London 2012 Olympic games ran smoothly found that fast moving people in corridors and tunnels become slow, shuffling, impatient and grumpy when the crowd density reached four people per square metre. At that point people just bump into each other and the 'flow' drops dramatically. They also found that people approaching a narrow doorway moved fastest along the edges of the tunnels rather than in the middle. This is surprising as water moves fastest in the middle and is very slow near the pipe boundary. So, in order to get people through a doorway faster they created a barrier (a circular pillar in the middle of the tunnel) just upstream of the opening. This split the flow into two and created two additional 'edges'. Rather than hindering the flow, this actually increased the flow. So next time you are in a slow moving tunnel, do not stay in the middle of the flow; always move to the side and you will get through faster.

3.7.2 How aeroplanes fly

Although, some people think that aircraft are lifted much in the same way as a flat stone is lifted as it skips across water (see Section 3.13), this is not the way it works. Aircraft rely on the energy equation to fly. An aircraft wing is specially shaped so that the airflow path is longer over the wing than under it (Figure 3.11a). So when an aircraft is taking off, the air moves faster over the wing than under it. This is necessary to maintain continuity of airflow around the wing. The result is an increase in kinetic energy over the wing. But the total energy around the wing does not change and so there is a corresponding reduction in the pressure energy above the wing. This means that the pressure above the wing is less than that below it and so the wing experiences a lift force. This can be a significant force; and as we witness everyday it can lift hundreds of tons of aeroplane into the air. It never ceases to amaze people and it works every time. Have you noticed that aeroplanes usually take-off into the wind? This is because the extra wind velocity increases the kinetic energy and provides extra lift. This is particularly important at take-off when an aeroplane is carrying its full fuel load and is at its heaviest.

The same principle is used in reverse on racing cars. In this case, the wing is upside down and located on the back of the car. The velocity of the air

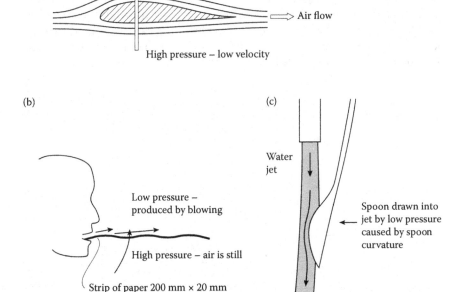

Figure 3.11 How aeroplanes fly. (a) Airflow around an aircraft wing. (b) Practical demonstration of lift. (c) Feeling the force with a spoon.

flowing over it, due to the forward movement of the car, produces a downward thrust which holds the car firmly on the road. The faster the car the greater is the down thrust which improves road holding and helps drivers to maintain high speeds even when cornering.

You can demonstrate this lift force yourself (Figure 3.11b). Tear off a strip of paper approximately 20 mm wide and 200 mm long. Grip the paper firmly in your teeth and blow gently across the top of the paper. You will see that it rises to a horizontal position. The blowing action increases the velocity of the air and hence reduces the pressure. The pressure of the still air below the paper is higher than above it and so the paper lifts – just like the aeroplane.

One way to feel the substantial force involved is to hold a spoon with its convex side close to water running from a tap (Figure 3.11c). Surprisingly, the water does not push the spoon away, rather it draws it into the water. This is because the water velocity increases as it flows around the spoon causing a drop in the pressure. This draws the spoon into the jet with surprising force. It is also the reason for that unpleasant feeling when you have a shower and the cold plastic curtain seems to stick to your body. It is all about pressure and velocity changes.

3.7.3 Airflow between buildings

Most people have noticed how suddenly the wind becomes much stronger in the gaps between buildings (Figure 3.12). This is another example of the effect of changing energy. A narrow gap causes an increase in wind velocity and a corresponding drop in air pressure. The pressure drop can cause doors to bang because the pressure between the buildings is lower than the pressure inside them (remember the air inside is still and at normal atmospheric pressure).

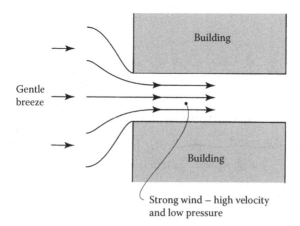

Figure 3.12 Applying the energy equation to airflow between buildings.

3.7.4 Fluid injectors

Farmers use fluid injectors to inject fertiliser into irrigation systems. Some use a pump which pushes the fertiliser into the pipe. But some are cleverer than this and can inject fluid without using additional energy. A short, narrow section of pipe (a venturi) is located in the main irrigation pipeline which causes the velocity to increase and the pressure to drop. Upstream of the venturi (where the pressure is high), a small diameter pipe takes off some of the flow and passes it through a fertiliser tank. A second pipe takes the mixture of water and fertiliser back to the pipeline where it is connected to the venturi section (where the pressure is low). So, the pressure drop caused by the change in pipe diameter drives the injector. The turbulence just downstream of the venturi, where the pipe expands again to its original size, ensures that the fertiliser is well mixed in the flow.

3.7.5 A very useful application

One very useful application of the energy equation is for measuring discharge. Changing the energy in pipes and channels produces changes in pressure which can be more easily measured than velocity. Using the energy and continuity equations, the pressure change is used to calculate velocities and hence discharges. (see Sections 4.10 and 6.7).

3.8 MOMENTUM

The momentum equation is the third tool in the box. Momentum is about movement and the forces which cause it (see Section 1.10). It is the link between force, mass and velocity and is used to calculate the forces created by water as it moves through pipes, hydraulic structures and machines like pumps and turbines.

The momentum equation is normally written as

force (N) = mass flow (kg/s) × change in velocity (m/s).

Mass flow can also be written as

mass flow (kg/s) = mass density (kg/m^3) × discharge (m^3/s)

$$= \rho Q.$$

And

velocity change $= v_2 - v_1,$

where v_1 and v_2 represent two velocities in a system. Put all these into the momentum equation

$$\text{Force } (F) = \rho Q \ (v_2 - v_1).$$

This is now in a form that is useful for calculating forces in hydraulics. An example of the use of this equation is shown in Box 3.6.

BOX 3.6   EXAMPLE: CALCULATING THE FORCE ON A PLATE FROM A JET OF WATER

A jet of water of diameter 60 mm and a velocity of 5 m/s hits a vertical plate. Calculate the force of impact of the jet on the plate (Figure 3.13).
    Remember, when dealing with momentum:

- Forces and velocities are vectors and so their direction is important as well as their magnitude.
- The force of the water jet on the plate is equal to the force of the plate on the water. They are the same magnitude but in opposite directions (remember Newton's third law).

Use the momentum equation to calculate the force on the plate

$$- F = \rho Q \ (v_2 - v_1).$$

Notice that flow and forces from left to right are assumed to be in a positive direction and so those which are from right to left are negative. F is the force of the plate on the water and is in the opposite direction to the flow, and so it is negative (working out the right direction can be rather tricky sometimes and so working with the momentum equation does take some practice).
    Reversing all the signs in the above equation makes F positive

$$F = \rho Q \ (v_1 - v_2).$$

The next step is to calculate the discharge Q

$$Q = va = v \times \frac{\pi d^2}{4}$$

$$= 5 \times \frac{\pi 0.06^2}{4} = 0.014 \, \text{m}^3/\text{s}$$

For this problem $v_2 = 0$ because the velocity of the jet after impact *in the direction of the flow* is zero. So, putting in the known values into the momentum equation

$$F = 1{,}000 \times 0.014 \times (5 - 0)$$
$$F = 70\,\text{N}$$

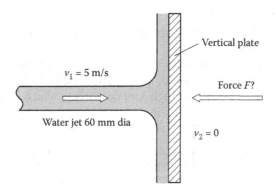

Figure 3.13 Applying momentum.

3.9 REAL FLUIDS

The assumption made so far in this chapter is that water is an ideal fluid. This means it has no viscosity and there is no friction between the flow and the boundaries, such as the inside of a pipe or the sides and bed of a channel. Water is a real fluid but its viscosity is low and so ignoring this has little or no effect on the design of pipes and channels. However, the friction between the flow and the boundary is important and cannot be ignored for design purposes. In fact, this is an important part of the design. We use a modified version of the energy equation to take account of this.

3.9.1 Taking account of energy losses

When water flows along pipes and channels, energy is lost from friction between the water and its boundaries, and we can account for this in the energy equation. Writing the energy equation for points 1 and 2 along a pipeline carrying a real fluid needs an additional term h_f to describe the energy loss between them

$$\frac{p_1}{\rho g} + \frac{v_1^2}{2g} + z_1 = \frac{p_2}{\rho g} + \frac{v_2^2}{2g} + z_2 + h_f$$

h_f is the most important element in this equation for determining the size of pipe or channel needed to carry a given flow. The question is how to measure or calculate it and what factors influence its magnitude. This was the challenge the nineteenth-century scientists faced, investigating fluid flow and the results of their work now form the basis of all pipe and channel design procedures. But more about this in Chapters 4 and 5.

3.9.2 Cavitation

Real fluids suffer from cavitation and it can cause lots of problems, particularly in pumps and control valves. It occurs when a fluid is moving very fast and as a consequence the pressure drops to very low values approaching zero (vacuum pressure).

The control valve on a pipeline provides a good example (Figure 3.14a). When the valve is almost closed, the water velocity under the gate can be very high. This also means high kinetic energy, which is gained at the

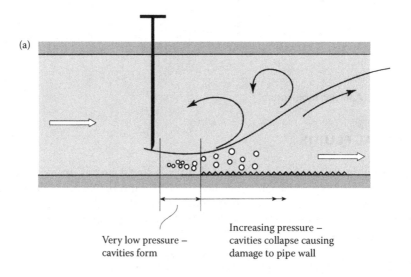

Very low pressure – cavities form

Increasing pressure – cavities collapse causing damage to pipe wall

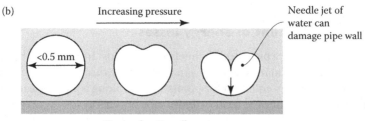

Stages of cavity collapse

Figure 3.14 Cavitation. (a) Cavitation under a sluice gate. (b) How cavitation bubbles collapse.

expense of the pressure energy. If the pressure drops below the vapour pressure of water (this is approximately 0.3 m absolute), bubbles called *cavities* start to form in the water. They are very small (less than 0.5 mm in diameter) but there are thousands of them and they give water a milky appearance. The bubbles are filled with water vapour and the pressure inside them is very low. But as the bubbles move under the gate and into the pipe downstream, the velocity slows, the pressure rises and the bubbles begin to collapse. It is at this point that the danger arises. If the bubbles collapse in the main flow they do no harm, but if they are close to the pipe wall they can do a great deal of damage. Notice the way in which the bubbles collapse (Figure 3.14b). As the bubble becomes unstable, a tiny needle jet of water rushes across the cavity and it is this which can do great damage even to steel and concrete because the pressure under the jet can be as high as 4,000 bar! (see Section 7.4.4 for more details of cavitation in pumps).

Some people confuse cavitation with air entrainment, but it is a very different phenomenon. Air entrainment occurs when there is turbulence at hydraulic structures and air bubbles are drawn into the flow. The milky appearance of the water is similar but the bubbles filled with air will not harm pumps and valves. Indeed, they can act as a cushion and protect structures from damage.

3.9.3 Boundary layers

Friction between water flow and its boundaries and the internal friction (viscosity) within the water gives rise to an effect known as the *boundary layer*. Water flowing in a pipe moves faster in the middle of the pipe than near the pipe wall. This is because friction between the water and the pipe wall slows down the flow. Very near to the pipe wall, water actually sticks to it and the velocity is zero, although it is not possible to see this with the naked eye. Gradually, the velocity increases further away from the wall until it reaches its maximum velocity in the centre of the pipe. To understand how this happens, imagine the flow is like a set of thin 'plates' that can slide over each other. The plate nearest to the wall is not moving and hence it tries to slow down to the plate next to it – the friction between the plates comes from the viscosity of the water. Plates further away from the wall are less affected by the boundary and so they move faster until the ones in the middle of the flow are moving fastest. All the flow affected by the pipe wall in this way is called the *boundary layer*. However, the use of the word *layer* can be misleading and is often confused with the layer of water closest to the pipe wall. My analogy with the sliding plates also does not help! So just to be clear, the boundary layer refers to all the flow which is affected by friction with the boundary. In the case of a pipe, the influence of the boundary is felt across the entire flow.

A graphical representation of the changes in velocity near a boundary is called the *velocity profile* (Figure 3.15a). The velocity changes from

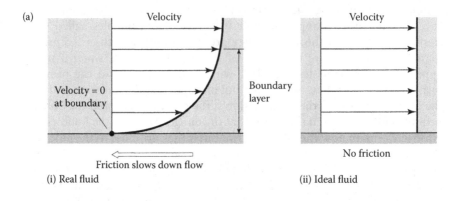

(i) Real fluid   (ii) Ideal fluid

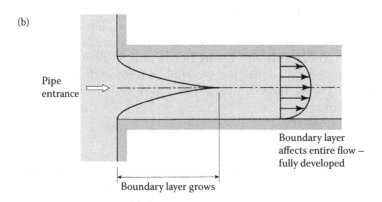

Figure 3.15 Boundary effects. (a) Velocity profiles. (b) Boundary layer growth.

zero near the boundary to a maximum in the centre of a pipe or channel. Compare this with the velocity profile for an ideal fluid. There is no viscosity and no boundary friction and so the velocity is the same across the entire flow.

Boundary layers grow as water enters a pipeline (Figure 3.15b). They quickly develop over the first few metres until they meet in the middle. From this point onwards the pipe boundary influences the entire flow in the pipe. In channels, the boundary effects of the bed and sides similarly grow over a few metres of channel and soon influence the entire flow. When the boundary layer fills the entire flow it is said to be *fully developed*. This fully developed state is the basis on which all pipe and channel formulae are based in Chapters 4 and 5.

3.9.3.1 The Earth's boundary layer

When the wind blows across the Earth's surface, it produces a boundary layer (Figure 3.16a). The wind is much slower near the ground where it is

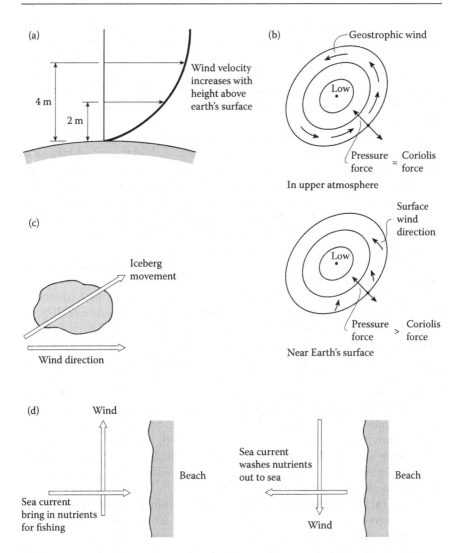

Figure 3.16 Earth's boundary layer. (a) When the wind blows. (b) Geostrophic wind and the Ekman spiral. (c) How icebergs move. (d) Upwelling.

most affected by friction between the air and the Earth's surface. Its influence extends many metres above the Earth's surface. For this reason, it is important to specify the height at which wind speed is measured in meteorological stations. At 2 m above the ground, the wind is much slower than at 4 m.

An interesting feature of the Earth's boundary layer is that not only does the wind slow down near the Earth's surface but it also gradually changes direction (Figure 3.16b). In the upper atmosphere, well beyond the boundary layer, the isobars (the lines of equal pressure) in a depression circle around the point of lowest pressure and the direction of the wind is always parallel to the isobars. This is because there is a balance between two important forces; the *Coriolis force,* which is a small but significant force that comes from the Earth's rotation, and the force trying to pull the air into the centre of the depression because of the difference in pressure. So, the wind circulates around the centre of the depression and is known as the *Geostrophic Wind.*

The Coriolis force does not affect us as individuals as we are too small but it does affect the movement of large masses, such as the air and the sea. Nearer the Earth's surface, in the boundary layer, the wind slows down and this reduces the effect of the Coriolis force. The two forces are now out of balance and hence the wind direction gradually changes as it is pulled in towards the centre of the depression. This is why the ground surface wind direction on weather maps is always at an angle to the isobars and pointing inwards towards the centre of the depression. This gradual twisting of the wind direction produces a spiral, rather like a spiral staircase, which is called the *Ekman Spiral.*

Vagn Walfrid Ekman (1874–1954), a Swedish scientist, first observed this spiral at sea. He noticed that in a strong wind, icebergs do not drift in the same direction as the wind but at an angle to it (Figure 3.16c). Surface winds can cause strong seawater currents and although the surface current may be in the direction of the wind, those currents below the surface are influenced both by the boundary resistance from the sea bed and the Coriolis force from the Earth's rotation. The effect is similar to that in the atmosphere. The lower currents slow down because of friction and gradually turn under the influence of the Coriolis force. So at the sea surface, the water is moving in the same direction as the wind, but close to the sea bed it is moving at an angle to the wind. As icebergs float over 90% submerged, their movement follows the water current rather than the wind direction and so they move at an angle to the wind.

This spiral effect is vital to several fishing communities around the world and is referred to as *up-welling* (Figure 3.16d). In Peru when the surface wind blows along the coast line, the boundary layer and the Coriolis force conspire to induce a current along the sea bed at right angles to the wind direction. This brings all the vegetative debris and plankton, on which fish like to feed, into the shallow waters of the shoreline and so the fishing is very good. However, when the wind blows in the opposite direction, the current is reversed and all the food is washed out to sea leaving the shallow coastal fishing grounds bare and the fishing industry devastated.

3.10 DRAG FORCES

Boundary layers occur around all kinds of objects, such as water flow around ships and submarines, airflow around aircraft and balls thrown through the air. Friction between the object and the fluid slows them down and it is referred to as a *drag force*. You can feel this force by putting your hand through the window of a moving car or in a stream of flowing water.

Sir George Stokes (1819–1903), an eminent physicist, was one of the first people to investigate drag by examining the forces on spheres falling through different fluids. He noticed that the spheres fell at different rates, not only because of the viscosity of the fluids but also because of the size of the spheres. He also found that the falling spheres eventually reach a constant velocity which he called the *terminal velocity*. This occurred when the force of gravity causing the balls to accelerate was balanced by the resistance resulting from the fluid viscosity and the size of the balls.

Stokes also demonstrated that for any object dropped in a fluid (or a stationary object placed in a flowing fluid which is essentially the same) there were two types of drag: *surface drag* or *skin friction*, which resulted from friction between the fluid and the object, and *form drag* which resulted from the shape and size of the object.

Water flowing around a bridge pier in a river provides a good example of the two types of drag. When the velocity is very low, the flow moves around the pier as shown in Figure 3.17a. The water clings to the pier and in this situation there is only surface drag and the shape of the pier has no effect. The flow pattern behind the pier is the same as the pattern upstream. But as the velocity increases, the boundary layer grows and the flow can no longer cling to the pier and so it *separates* (Figure 3.17b). It behaves like a car that is travelling too fast to get around a tight bend. It spins away from the pier and creates several small whirl pools which are swept downstream. These are called *vortices* or *eddies,* and together they form what is known as the *wake,* which gets wider as it gradually draws in more of the river flow through friction (Figure 3.17b). The flow pattern behind the pier is now quite different from that in front and in the wake the pressure is much lower than in front. It is this difference in pressure that produces the *form drag.* It is additional to the *surface drag* and its magnitude depends on the shape of the pier. Going back to your hand through the car window, notice how the force changes when you place the back or side of your hand in the direction of the flow. The shape of your hand in the flow determines the form drag.

Form drag is usually more important than surface drag and it can be reduced by shaping a bridge pier so that the water flows around it more easily and separation is delayed or avoided. Indeed, if separation could be avoided completely then form drag would be eliminated and the only concern would be surface drag. Shaping piers to produce a narrow wake and reduce form drag is often called *streamlining* (Figure 3.17c). This is the basis of design not just for bridge piers but also for aircraft, ships and cars – to reduce drag and so increase the speed or reduce the energy requirements. As you might imagine the study of drag is now a very sophisticated science as manufacturers develop computer modelling and use wind tunnels to get the best designs.

(a)   Very low velocity

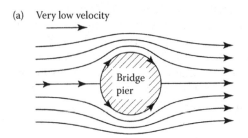

(b)

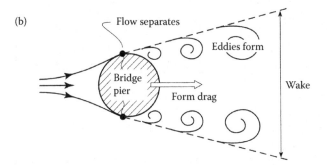

(c)

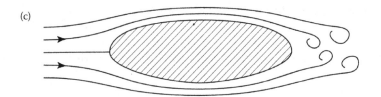

(d)

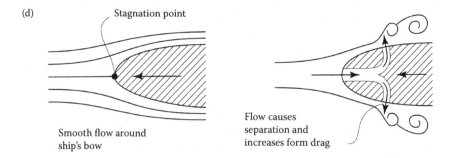

Figure 3.17 Boundaries and drag. (a) Surface drag only – no form drag. (b) Increasing velocity causes separation to occur. (c) Form drag reduced by streamlining. (d) Using form drag to stop tankers.

Swimmers too can benefit from reducing drag. This is particularly important at competitive levels when a few hundredths of a second can mean the difference between a gold and a silver medal. Approximately, 90% of the drag on a swimmer is form drag, only 10% is surface drag. Some female swimmers try to reduce form drag by squeezing into a swimsuit 2 or 3 sizes too small for them in order to improve their shape in the water.

Although women swimmers may seem to have an advantage in having a more streamline shape than bulky males, their shape does present some hydraulic problems. A woman's breasts cause early flow separation which increases turbulence and form drag. One swimwear manufacturer has found a solution to this by using a technique used by the aircraft industry to solve a similar problem. Aircraft wings often have small vertical spikes on their upper surface and these stop the flow from separating too early by creating small vortices, that is, zones of low pressure close to the wing surface. This not only reduces form drag significantly but also helps to avoid stalling (very early separation), which can be disastrous for an aircraft. The new swimsuit has tiny vortex generators located just below the breasts which cause the boundary layer to cling to the swimmer. This stops the boundary layer from separating and hence reduces form drag. The same manufacturer has also developed a ribbed swimsuit, which creates similar vortices all along the swimmer's body to try and stop the flow from separating. The manufacturer claims a 9% reduction in drag for the average female swimmer over a conventional swimsuit.

Dolphins probably have the answer for reducing drag. They are well known for their natural shape and skin for swimming. Both their form and surface drag are very low, which enables them to move through the water with incredible ease and speed – something that human beings have been trying to emulate for many years!

There is a way of calculating drag force

$$\text{drag force} = \frac{1}{2}C\rho a v^2$$

where ρ is fluid density (kg/m^3), a is the cross-sectional area (m^2), v is velocity (m/s) and C is drag coefficient. The coefficient C is dependent on the shape of the body, the flow velocity and the fluid density.

3.10.1 Stopping supertankers

Supertankers, because of their enormous size, are designed with low drag in mind so they can travel the seas with only modest energy requirements to drive them. The problem comes when they want to stop. When the engines stop, they can travel for several kilometres before drag forces finally stop them. How then do you put on the brakes on a supertanker? One way is to increase the ship's form drag by taking advantage of the stagnation point at the bow of the ship to push water through an inlet pipe in the bow and out at the sides of the ship (Figure 3.17d). This flow at right angles to the movement of the ship causes the boundary layer to separate and greatly increase the form drag. It is as if the ship is suddenly made much wider and this upsets its streamline shape.

3.11 EDDY SHEDDING

Eddies which form in the wake around bridge piers can also cause other problems besides drag. Eddies are not shed from each side of the pier at the same time but alternately, first from the one side, then from the other. Under the right flow conditions, large eddies can form and the *alternate eddy shedding* induces a sideways force which pushes the pier from side to side in a slow rhythmic vibration (Figure 3.18a). This problem is not just confined to bridge piers. It is a problem for tall chimneys and for bridge decks in windy conditions. The vibration can become so bad that structures collapse.

A famous suspension bridge, the Tacoma Narrows Bridge in the United States, was destroyed in the 1930s because of this problem (Figure 3.18b). In order to protect traffic from high winds blowing down the river channel, the sides of the bridge were boarded up. Unfortunately, the boarding deflected the wind around the bridge deck, the airflow separated forming large eddies, and this set the bridge deck oscillating violently up and down. The bridge deck was quite flexible as it was a suspension bridge and could in fact tolerate quite a lot of movement but this was so violent that eventually it destroyed the bridge.

The solution to the problem was quite simple. If the side panels had been removed, this would have stopped the large eddies from forming and there would have been no vibration. So, next time you are on a suspension bridge and a strong wind is blowing and you are feeling uncomfortable be thankful that the engineers have decided not to protect you from the wind by boarding up the sides.

A similar problem can occur around tall chimneys when eddies are shed in windy conditions. To avoid large eddies forming, a perforated sleeve or a spiral collar is placed around the top of the chimney. This breaks up the flow into lots of small eddies which are usually quite harmless.

3.12 MAKING BALLS SWING

Sports players soon learn how useful boundary layers can be when they realise that balls can be made to move in a curved path through the air and so confuse their opponents. A good example of this is the way some bowlers are able to make a ball 'swing' (move in curved path) in cricket.

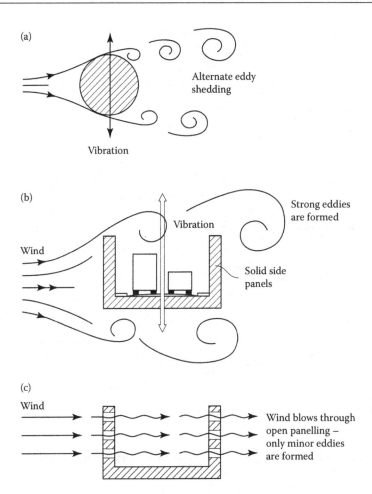

Figure 3.18 Eddy shedding problems. (a) Alternate eddy shedding around a cylinder. (b) Flow around a bridge deck with solid sides. (c) With open panels.

When a ball is bowled (for non-cricket enthusiasts this means throw), the air flows around it and at some point it separates (Figure 3.19). When the separation occurs at the same point all around the ball then it moves along a straight path. However, when it occurs asymmetrically, there is a larger pressure on one side of the ball and so it starts to move in a curved path (i.e. it swings). The bowler's task is to work out how to do this.

Laboratory experiments have shown that as the air flows around a ball, it can be either turbulent or laminar (these are two different kinds of flow described in Section 4.3.1). When it is turbulent, the air clings to the ball more easily than when it is laminar. So, the bowler tries to make the airflow turbulent on one side of the ball and laminar on the other. This is done by making one side very smooth and the other side rough. In cricket, this

(a)

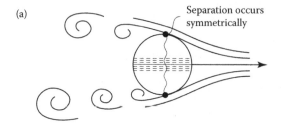

Separation occurs
symmetrically

(b)

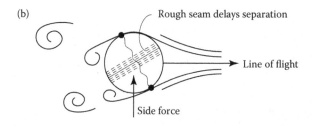

Rough seam delays separation

Line of flight

Side force

(c)

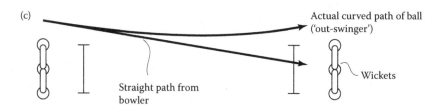

Actual curved path of ball
('out-swinger')

Wickets

Straight path from
bowler

Figure 3.19 Making balls swing. (a) Symmetrical flow around a smooth ball. (b) Asymmet-
ric flow causes ball to swing. (c) High velocity produces an 'out-swinger'.

situation is helped by a special stitched seam around the middle of the ball
which ensures that the ball is rough enough to create turbulent conditions.
The ball is bowled so that the polished side of the ball is facing the batsman and the seam is at an angle to the main direction of travel. The airflow
on the smooth side separates earlier than on the rough side and so the
ball swings towards the turbulent side. The cross force can be up to 0.8 N
depending on how fast the ball is travelling and may cause a swing of up to
0.6 m in 20 m, which is the length of a cricket pitch. This can be more than
enough to seriously confuse a batsman who may be expecting a straight
ball. This is why bowlers seem to spend so much of their time polishing the
ball on their trousers prior to their run-up in order to get it as smooth as
possible to get the maximum swing. The swing may be in or out depending
on how the bowler holds the ball. However, not all the surprise is with the
bowler. An observant batsman may know what is coming by looking to see
how the bowler is holding the ball and so anticipate the swing.

Sometimes, strange things happen which even puzzle those who understand hydraulics. Just occasionally, bowlers have noticed that a ball that was meant to swing in towards the batsman swings away from him instead – an *outswinger*. What happens is that when a ball is bowled fast enough, the entire airflow around the ball turns turbulent and so the separation occurs much earlier than for laminar flow. The stitched seam around the ball now acts as a ramp causing the air to be pushed away, creating a side force in the opposite direction to what was expected. This causes great delight for the bowler but it can give the batsman quite a fright. But most batsmen can relax as this special swing only occurs when the ball reaches 130–150 km/h and only a few bowlers can actually reach this velocity. However, some unscrupulous bowlers have discovered a way of doing this at much lower velocities. By deliberately roughening the ball on the one side (which is not allowed) and polishing it on the other (which is allowed), they can bowl an out swinger at much lower velocities. This caused a major row in cricket in the early 1990s and again in 2006 when a Pakistani bowler was accused of deliberately roughening the ball. Imran Khan though was famous for his high-speed bowling and could produce outswingers without resorting to such tactics. It is of course allowed for the ball to scuff or become rough naturally through play but this can take some time.

Causing a ball to spin at the same time as driving it forward can also add to the complexities of airflow and also to the excitement of ball sports. Some famous ball swings in the recent years resulted in the goals scored by the Brazilian footballer, Roberto Carlos in 1997 and by David Beckham in the 2006 World Cup games. In each case, the goal area was completely blocked by the opposing team players. As each player kicked the ball, it seemed to be heading for the corner flag but instead it followed a curved path around the defending players and into the goal. They achieved this amazing feat by striking the ball on its edge causing it to spin, which induced a sideways force. This, together with the boundary layer effect and a great deal of skill (and a little luck) produced some of the best goals ever scored. Since those early days of Beckham and Carlos, lots of players have now mastered the skills, even children at school seem able to copy the skills as well.

3.13 SUCCESSFUL STONE-SKIPPING

Skipping stones across water has been a popular pastime for many thousands of years (Figure 3.20). Apparently, the Greeks started it and according the *Guinness Book of Records* the world record is held by Kurt Steiner. It was set in 2003 at 40 rebounds.

Various parameters affect the number of skips such as the shape, weight, velocity and spin. The stone should ideally be like a small, flat plate. This was very much an art until research undertaken at the Institut de Recherche sur les Phénomènes Hors Equilibre in Marseille and published in *Nature* in 2004

Figure 3.20 The author prepares for an attempt on the stone-skipping record.

investigated the physics of this idle pastime. Among the many parameters investigated, the most important one was the angle at which the stone hits the water. The 'magic' angle, as the researchers described it, was 20° and at this angle the energy dissipated by the stone impact with the water is minimised. So at this angle, you will achieve the maximum number of skips. Spinning the stone also helps because this stabilises the stone owing to the gyroscopic effect. You may not reach 40 skips but at this angle you have the best chance.

Interestingly, stone skipping is often thought to provide a theory for lift on an aircraft wing. Instead of water hitting the underside of a stone and lifting it, the idea of air hitting the underside of an aerofoil has some appeal. But it is quite wrong. At a wing angle of 20°, an aircraft would be in danger of stalling as the flow above the wing would separate. The wing lift comes from the energy changes which take place around the wing (see Section 3.7.2).

3.14 SOME EXAMPLES TO TEST YOUR UNDERSTANDING

1. To measure the discharge in a pipe, a 10 L bucket is used to catch the flow at its outlet. If it takes 3.5 s to fill the bucket, calculate the discharge in m^3/s. Calculate the velocity in the pipe when the diameter is 100 mm (0.35 m/s; 0.0028 m^3/s).

2. A main pipeline 300 mm in diameter is carrying a discharge of 0.16 m³/s and a smaller pipe of diameter 100 mm is joined to it to form a tee junction and takes 0.04 m³/s. Calculate the velocity in the 100-mm pipe and the discharge and velocity in the main pipe downstream of the junction (5.09 m/s; 0.12 m³/s; 1.7 m/s).

3. A fountain is to be designed for a local park. A nozzle diameter of 50 mm is chosen and the water velocity at the nozzle will be 8.5 m/s. Calculate the height to which the water will rise. The jet passes through a circular opening 2 m above the nozzle. Calculate the diameter of the opening so that the jet just passes through without interference (3.68 m; opening greater than 61 mm).

4. A pipeline 500 mm diameter is carrying a discharge of 0.5 m³/s at a pressure of 55 kN/m² reduces to 300 mm diameter. Calculate the velocity and pressure in the 300 mm pipe (7.14 m/s; 33 kN/m²).

Chapter 4

Pipes

4.1 INTRODUCTION

Pipes are a common feature of water supply systems and have many advantages over open channels. They are completely enclosed, usually circular in section and always flow full of water. This is in contrast to channels which are open to the atmosphere and can have many different shapes and sizes – but more about channels in Chapter 5. One big advantage of pipes is that water can flow uphill as well as downhill, and so land topography is not such a constraint when taking water from one location to another.

There are occasions when pipes do not flow full – one example is gravity flow sewers (see Section 8.4). They take sewage away from homes and factories and often only flow partially full under the force of gravity in order to avoid pumping. They look like pipes and are indeed pipes but hydraulically they behave like open channels. The reason pipes are used for this purpose is that sewers are usually buried below ground to avoid public health problems and it would be difficult to bury an open channel.

4.2 TYPICAL PIPE FLOW PROBLEM

Pipe flow problems usually involve calculating the right size of pipe to use for a given discharge. A typical example is a water supply to a village (Figure 4.1). A pipeline connects the main storage reservoir to a small service (storage) tank just outside the village which then supplies water to individual houses. The required discharge (Q m³/s) for the village is determined by the water demand of each user and the number of users being supplied (see Section 8.3). We now need to determine the right size of pipe to use.

A formula to calculate pipe size would be ideal. However, to get there we first need to look at the energy available to 'push' water through the system, so the place to start is the energy equation. But this is a real fluid problem and hence energy losses due to friction must be taken into account. So writing the energy equation for two points in this system – point 1 is at

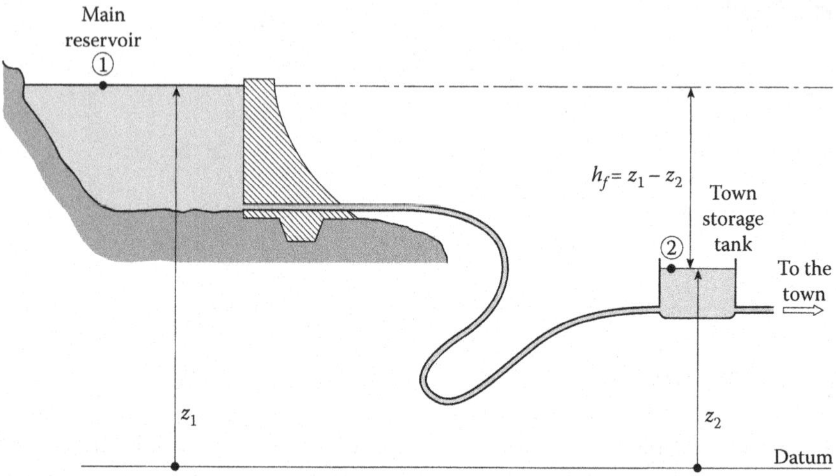

Figure 4.1 Typical pipe flow problem.

the main reservoir and point 2 is at the service tank – and allowing for the energy loss as water flows between the two

$$\frac{p_1}{\rho g} + \frac{v_1^2}{2g} + z_1 = \frac{p_2}{\rho g} + \frac{v_2^2}{2g} + z_2 + h_f.$$

Points 1 and 2 are carefully chosen in order to simplify the equation and also the solution. Point 1 is at the surface of the main reservoir where the pressure p_1 is atmospheric pressure and so is equal to zero (remember we are working in gauge pressures). Point 2 is also at the water surface in the service tank and so p_2 is zero as well. The water velocities v_1 and v_2 in the reservoir and the tank are very small and hence the kinetic energy terms are also very small and can be assumed to be zero. This leaves just the potential energy terms z_1 and z_2 and the energy loss term h_f. So the energy equation simplifies down to

$$h_f = z_1 - z_2.$$

$z_1 - z_2$ is the difference in water levels between the reservoir and the storage tank and this represents the energy available to 'push' water through the system. h_f is the energy loss due to friction in the pipe. The energy available is usually known and hence this means we also know the amount of energy that can be lost through friction. The question now is: Is there a link between energy loss h_f and the pipe diameter? The short answer is yes – but it has taken some 150 years of research to sort this out. So, let us first step through a bit of history and see what it tells us about pipe flow.

4.3 FORMULA TO LINK ENERGY LOSS AND PIPE SIZE

Some of the early research work on pipe friction was done by Osborne Reynolds (1842–1912), a mathematician and engineer working at the University in Manchester in the United Kingdom. He measured the pressure loss in pipes of different lengths and diameters at different discharges with some interesting results. At low flows he found that the energy loss varied directly with the velocity. So when the velocity was doubled, the energy loss also doubled. But at high flows the energy loss varied as the square of the velocity. So, when the velocity was doubled, the energy loss increased fourfold. Clearly, Reynolds was observing two quite different types of flow. This thinking led to Reynolds classic experiment that established the difference between what we now call *laminar* and *turbulent* flow, and formulae which would enable the energy loss to be calculated for each flow type from the knowledge of the pipes themselves.

4.3.1 Laminar and turbulent flow

Reynolds experiment involved setting up a glass tube through which he could pass water at different velocities (Figure 4.2). A thin jet of coloured dye was injected into the flow so that the flow patterns were visible.

When the water moved slowly, the dye remained in a thin line as it followed the flow path of the water down the pipe. This was described as *laminar flow*. It was as though the water was moving like a set of thin layers, like a pack of cards, each card sliding over the others, and the dye injected between two of the layers. This type of flow rarely exists in nature and hence is not of great practical concern to engineers. However, you can see it occasionally under very special conditions. Examples include smoke rising in a thin column from a chimney on a very still day or a slow flow of water from a tap that looks so much like a glass rod that you feel you could get hold of it. Blood flow in our bodies is usually laminar.

The second and more common type of flow he identified was *turbulent flow*. This occurred when water was moving faster. The dye was broken up as the water whirled around in a random manner and was dissipated throughout the flow. Turbulence was a word introduced by Lord Kelvin (1824–1907) to describe this kind of flow behaviour.

There are very clear visual differences between laminar and turbulent flow but what was not clear was how to predict which one would occur in any given set of circumstances. Velocity was obviously important. As velocity increased, so the flow would change from laminar to turbulent flow. But it was obvious from the experiments that velocity was not the only factor. It was Reynolds who first suggested that the type of flow depended not only on velocity (v) but also on mass density (ρ), viscosity (μ) and pipe diameter (d). He put these factors together in a way which is now called the *Reynolds Number* (R_e) in recognition of his work.

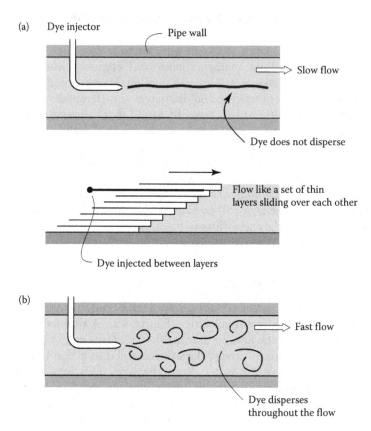

Figure 4.2 Laminar and turbulent flow. (a) Laminar flow. (b) Turbulent flow.

$$\text{Reynolds No } (R_e) = \frac{\rho v d}{\mu}.$$

Note that Reynolds Number has no dimensions. All the dimensions cancel out. Reynolds found that he could use this number to reliably predict when laminar and turbulent flow would occur.

$R_e < 2,000$ flow would always be laminar.
$R_e > 4,000$ flow would always be turbulent.

Between $R_e = 2,000$ and $4,000$, he observed a very unstable zone as the flow seemed to jump from laminar to turbulent and back again as if the flow could not decide which of the two conditions it preferred. This is a zone to avoid as both the pressure and flow fluctuate widely in an uncontrolled manner.

Reynolds Number also shows how important is viscosity in pipe flow. Low Reynolds Number ($R_e < 2,000$) means that viscosity (μ) is large compared with the term ρvd. So, viscosity is important in laminar flow and cannot be ignored. High Reynolds Number ($R_e > 4,000$) means viscosity is small compared with the ρvd term and so it follows that viscosity is less important in turbulent flow. This is the reason why engineers ignore the viscosity of water when designing pipes and channels as it has no material effect on the solution. Ignoring viscosity also greatly simplifies pipeline and channel design.

It has since been found that Reynolds Number is very useful in other ways besides telling us the difference between laminar and turbulent flow. It is used extensively in hydraulic modelling (physical models rather than mathematical models) for solving complex hydraulic problems. When a problem cannot be solved using some formula, another approach is to construct a small-scale model in a laboratory and test it to see how it performs. The guideline for modelling pipe systems (or indeed any fully enclosed system) is to ensure that the Reynolds Number in the model is similar to the Reynolds Number in the real situation. This ensures that the forces and velocities are similar (known as *dynamic similarity*) so that the model, as near as possible, produces similar results to those expected in the real pipe system.

Although it is useful to know that laminar flow exists, it is not important in practical hydraulics for designing pipes and channels and so only turbulent flow is considered in this text. Turbulent flow is very important to us in our daily lives. Indeed, it would be difficult for us to live if it was not for the mixing that takes place in turbulent flow which dilutes fluids. When we breathe out, the carbon dioxide from our lungs is dissipated into the surrounding air through turbulent mixing. If it did not disperse in this way, we would have to move our heads to avoid breathing in the same gases as we had just breathed out. Car exhaust fumes are dispersed in a similar way; otherwise, we could be quickly poisoned by the intake of concentrated carbon monoxide. Life could not really exist without turbulent mixing.

4.3.2 Formula for turbulent flow

Several formulae link energy loss with pipe size for turbulent flow but one of the most commonly used today is that devised by Julius Weisbach (1806–1871) and Henry Darcy (1803–1858). It is called the *Darcy–Weisbach equation* in recognition of their work

$$h_f = \frac{\lambda l v^2}{2gd},$$

where λ is a friction factor, l is pipe length (m), v is velocity (m/s), g is gravity constant (9.81 m/s^2) and d is pipe diameter (m). This formula shows that energy loss depends on pipe length, velocity and diameter and also on friction between the pipe and the flow as represented by λ:

- *Length* has a direct influence on energy loss. The longer the pipeline the greater the energy loss.
- *Velocity* has a great influence on energy loss because it is the square of the velocity that counts. When the velocity is doubled (say by increasing the discharge), the energy loss increases fourfold. It is usual practice in water supply systems to keep the velocity below 1.6 m/s. This is done primarily to avoid excessive energy losses but it also helps to reduce water hammer problems (see Section 4.14).
- *Pipe diameter* has the most dramatic effect on energy loss. As the pipe diameter is reduced, so the energy losses increase, not only because of the direct effect of d in the formula but also because of its effect on the velocity v (remember the discharge equation $Q = va$). The overall effect of reducing the diameter by half (say from 300 to 150 mm) is to increase h_f by 32 times (see Box 4.1).
- *Pipe friction* λ–unfortunately, this is not just a simple measure of pipe roughness; it depends on several other factors which are discussed in detail in the next section.

Take care when using the Darcy–Weisbach formula as some textbooks, particularly American, use f as the friction factor and not λ. They are not the same. The link between them is $\lambda = 4f$.

4.4 THE λ STORY

It would be convenient if λ was just a constant number for a given pipe that depended only on its roughness and hence its resistance to the flow. But few things are so simple and λ is no exception. Some of the earliest work on pipe friction was done by Paul Blazius in 1913. He carried out a wide range of experiments on different pipes and different flows, and came to the conclusion that λ depended only on the Reynolds Number, and surprisingly, the roughness of the pipe seemed to have no effect at all on friction.

From this he developed a formula for λ

$$\lambda = \frac{0.316}{R_e^{0.25}}.$$

Another investigator was Johann Nikuradse (1894–1979) working in Germany may well have been puzzled by the Blazius results. He set up a series of laboratory experiments in the 1930s with different pipe sizes and flows. He roughened the inside of the pipes with sand grains of a known

---

**BOX 4.1   EXAMPLE: HOW PIPE DIAMETER
AFFECTS ENERGY LOSS**

A pipeline 1,000 m long carries a flow of 100 L/s. Calculate the energy loss when the pipe diameter is 300, 250, 200 and 150 mm. Assume $\lambda = 0.04$.

The first step is to calculate the velocities for each pipe diameter using the discharge equation

$$Q = va.$$

And so

$$v = \frac{Q}{a}.$$

Use this equation to calculate velocity v for each diameter and then use the Darcy–Weisbach equation to calculate h_f. The results are shown in Table 4.1. Notice, the very large rise in head loss as the pipe diameter is reduced. Clearly, the choice of pipe diameter is a critical issue in any pipeline system.

Notice how rapidly the velocity and the energy losses increase as the pipe diameter decreases.

Table 4.1 Effect on head loss of changing pipe diameter

Diameter (m)	Pipe area (m²)	Velocity (m/s)	Head loss h_f (m)
0.30	0.07	1.43	13.6
0.25	0.049	2.04	33.3
0.20	0.031	3.22	103.7
0.15	0.018	5.55	418.6

---

size in order to create different but known roughness. His data showed that values of λ were independent of Reynolds Number and depended only on the roughness of the pipe. Clearly, either someone was wrong or they were both right and each was looking at something different.

4.4.1 Smooth and rough pipes

We now know that both investigators were right but they were looking at different aspects of the same problem. Blazius was looking at flows with relatively low Reynolds Numbers (4,000–100,000) and his results refer to what are now called *smooth pipes*. Nikuradse's experiments dealt with high Reynolds Number flows (greater than 100,000) and his results refer to what are now called *rough pipes*. Both Blazius and Nikuradse results are shown graphically in Figure 4.3a. This is a graph with a special logarithmic scale for Reynolds Number so that a wide range of values can be shown on the same graph. It shows how λ varies with both Reynolds Number and pipe roughness which is expressed as the height of the sand grains (k) divided by the pipe diameter (d). The Blazius formula produces a single line on this graph and is almost a straight line.

The terms rough and smooth refer as much to the flow conditions in pipes as to the pipes themselves, and so paradoxically, it is possible for

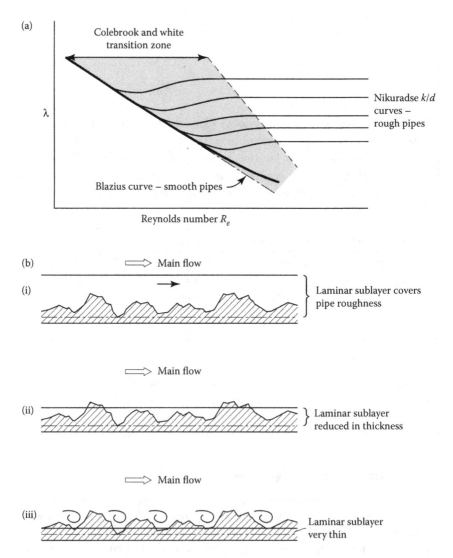

Figure 4.3 λ story. (a) Log graph of λ and Reynolds number. (b(i)) Smooth pipe flow. (b(ii)) Transitional pipe flow. (b(iii)) Rough pipe flow.

the same pipe to be described as both rough and smooth. Roughness and smoothness are also relative terms. How the inside of a pipe feels to touch is not a good guide to its smoothness in hydraulic terms. Pipes which are smooth to the touch can still be quite rough hydraulically. However, a pipe that feels rough to touch will be very rough hydraulically and very high energy losses can be expected.

As there are two distinct types of flow, it implies that there must be some point or zone where the flow changes from one to the other. This is indeed the case. It is not a specific point but a zone known as the *transition zone* when λ depends on both Reynolds Number and pipe roughness (Figure 4.3a). This zone was successfully investigated by CF Colebrook and CM White working at Imperial College in London in the 1930s and they developed a formula to cover this flow range. This is not quoted here as it is quite a complex formula and in practice there is no need to use the formula because it has now been simplified to design charts. These can be used to select pipe sizes for a wide range of hydraulic conditions. The use of typical pipe charts is described later in this chapter in Section 4.8.

The transition zone between smooth and rough pipe flow should *not* be confused with the transition zone from laminar to turbulent flow as is often done. The flow is fully turbulent for all smooth and rough pipes and the transition is from smooth to rough pipe flow.

To summarise the different flows in pipes:

Laminar flow

↓

Transition from laminar to turbulent flow (this zone is very unstable and should be avoided)

↓

Turbulent flow – smooth pipe flow

↓

Transition from smooth to rough pipe flow

↓

Rough pipe flow

4.4.1.1 Physical explanation

Since those early experiments, modern scientific experiments and equipment have enabled investigators to look more closely at what happens close to a pipe wall. This has resulted in a physical explanation for smooth and rough pipe flow (Figure 4.3b). Investigators have found that even when the flow is turbulent there exists a very thin layer of fluid, less than 1 mm thick, close to the boundary that is laminar. This is called the *laminar sub-layer*. At low Reynolds Numbers, the laminar sub-layer is at its thickest and completely covers the roughness of the pipe. The main flow is unaffected by the boundary roughness and is influenced only by viscosity within the laminar sub-layer. It seems that the layer covers the roughness like a blanket and protects the flow from the pipe wall. This is the *smooth pipe flow* that Blazius investigated. As Reynolds Number increases, the laminar sub-layer becomes thinner and roughness elements start to protrude into the main flow. The flow is now influenced both by viscosity and pipe roughness. This is the *transition zone*. As Reynolds Number is further increased the laminar sub-layer all but disappears and the roughness of the pipe wall takes over and dominates the friction. This is *rough pipe flow* which Nikuradse investigated.

Commercially manufactured pipes are not artificially roughened with sand like experimental pipes; they are manufactured as smooth as possible to reduce energy losses. For this reason, they tend to fall within the transition zone where λ varies with both Reynolds Number and pipe roughness.

4.5 HYDRAULIC GRADIENT

One way of showing energy losses in a pipeline is to show them diagrammatically (Figure 4.4a). The total energy line marked *e—e—e* shows how the total energy changes along the pipeline. As energy is lost from friction, the line is always downwards in the direction of the flow. It connects the water surfaces in the two tanks. There is little energy loss at the entrance to the pipeline but there is a bigger step at the downstream tank to represent the energy loss as water flows from the pipeline into the tank. The energy line is not necessarily parallel to the pipeline. The pipeline usually follows the natural ground surface profile.

Although total energy is of interest, pressure is more important because this determines how strong the pipes must be to avoid bursts. So the second line is the pressure line *h—h—h*. It is always below the energy line but parallel to it, to represent the pressure (pressure energy). This shows the pressure change along the pipeline. Imagine standpipes are attached to the pipe. Water would rise up to this line to represent the pressure head (Figure 4.4a). The difference between the two lines is the kinetic energy. Notice how both the energy line and the hydraulic gradient are straight lines. This shows that the rate of energy loss and the pressure loss are uniform (at the same rate). The slope of the pressure line is called the *hydraulic gradient* and is calculated as follows

$$\text{hydraulic gradient} = \frac{h_f}{l},$$

where h_f is change in pressure (m); and l is the pipe length over which the pressure change takes place (m). The hydraulic gradient has no dimensions as it comes from dividing a length in metres by a head difference in metres. It is expressed in terms of metres head per metre length of pipeline. As an

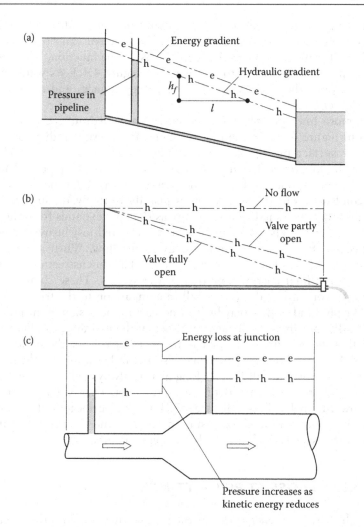

Figure 4.4 Hydraulic gradient. (a) Flow between two reservoirs. (b) Hydraulic gradient changes with flow. (c) Hydraulic gradient can rise and fall.

example, a hydraulic gradient of 0.02 means that for every 1 m of pipeline, there will be a pressure loss of 0.02 m. This may also be written as 0.02 m/m or as 2 m/100 m of pipeline. The latter reduces the number of decimal places and means that for every 100 m of pipeline, 2 m of head is lost through friction. So if a pipeline is 500 m long (there are 5–100 m lengths), the pressure loss over 500 m will be $5 \times 2 = 10$ m head.

The hydraulic gradient is not a fixed line for a pipe; it depends on the flow (Figure 4.4b). When there is no flow, the gradient is horizontal but when there is full flow the gradient is at its steepest. Adjusting the outlet valve will produce a range of gradients between these two extremes.

The energy gradient can only slope downwards in the direction of flow to show how energy is lost, but the hydraulic gradient can slope upwards as well as downwards. An example of this is a pipe junction, when water flows from a smaller pipe into a larger one (Figure 4.4c). As water enters the larger pipe, the velocity and kinetic energy reduces and the pressure energy increases.

Two more finer points about the energy and hydraulic gradients are shown in Figure 4.4a. At the first reservoir, the energy gradient starts at the water surface but the hydraulic gradient starts just below it. This is because the kinetic energy increases as water enters the pipe and hence there is a corresponding drop in the pressure energy. As the flow enters the second reservoir, the energy line is just above the water surface. This is because there is a small loss in energy as the flow expands from the pipe into the reservoir. The hydraulic gradient is located just below the water level because there is still kinetic energy in the flow. When it enters the reservoir, it changes back to pressure energy. The downstream water level represents the final energy condition in the system. These changes close to the reservoirs are really very small in comparison to the friction losses along the pipe and so they play little or no part in the design of the pipeline.

Normally, pipelines are located well below the hydraulic gradient. This means that the pressure in the pipe is always positive – see the standpipe in Figure 4.4a. Even though the pipe may rise and fall as it follows the natural ground profile, water will flow as long as it is always below the hydraulic gradient and provided the outlet is below the inlet. There are limits to how far below the hydraulic gradient a pipeline can be located. The further below, the higher will be the pressure in the pipe and the risk of a burst if the pressure exceeds the limits set by the pipe manufacturer.

4.6 ENERGY LOSS AT PIPE FITTINGS

Although there is an energy loss at the pipe connection with the reservoir in Figure 4.4a, this is usually very small in comparison with the loss in the main pipeline and hence it is often ignored. Similar losses occur at pipe bends, reducers, pipe junctions and valves; and although each one is small, together they can add up. They can all be calculated individually but normal design practice is to simply increase the energy loss in the main pipeline by 10% to allow for all these minor losses.

4.7 SIPHONS

Siphon is the name given to sections of pipe that rise above the hydraulic gradient. Normally, pipes are located well below the hydraulic gradient and this ensures that the pressure is always positive and so well above

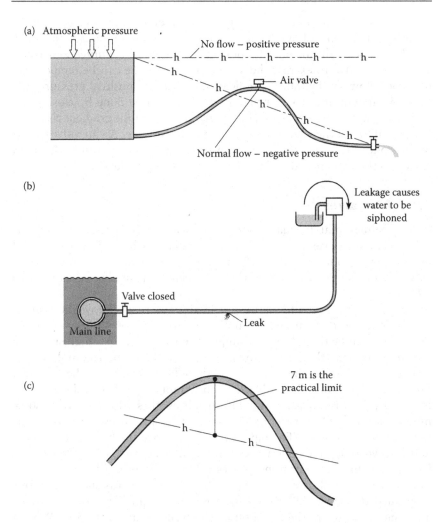

Figure 4.5 Siphons. (a) Typical siphon with pipe above hydraulic gradient. (b) Device for finding leaks. (c) Siphon limit.

atmospheric pressure. Under these conditions, water flows freely under gravity provided the outlet is lower than the inlet (Figure 4.4a). But when part of a pipeline is located above the hydraulic gradient, even though the outlet is located below the inlet, water will not flow without some help (Figure 4.5a). This is because the pressure in the section of the pipe above the hydraulic gradient is negative.

4.7.1 How they work

Before water will flow, all the air must be taken out of the pipe to create a vacuum. When this happens, the atmospheric pressure on the open water surface pushes water into the pipe to fill the vacuum and once it is full of water it will begin to flow. Under these conditions the pipe is working as a *siphon*. Taking the air out of a pipeline is known as *priming*. Sometimes a pump is needed to extract the air but if the pipeline can be temporarily brought below the hydraulic gradient, the resulting positive pressure will push the air out and it will prime itself. This can be done by closing the main valve at the end of the pipeline so that the hydraulic gradient rises to a horizontal line at the same level as the reservoir surface. An air valve on top of the siphon then releases the air. Once the pipe is full of water, the main valve can then be opened and the pipeline will flow normally.

4.7.2 Air valves

Even pipelines that normally operate under positive pressures have air valves. These release air which can accumulate at high spots. It is good practice to include air valves at such locations. They can be automatic valves or just simple gate valves that are opened manually occasionally to release air.

It can sometimes be difficult to spot an air valve that is above the hydraulic gradient and this can lead to problems. An engineer visiting a remote farm saw what he thought was a simple gated air valve on a high spot on a pipeline supplying the farm with water. Air does tend to accumulate over time and can restrict the flow. So, he thought he would do the farmer a favour and open the valve to bleed off any trapped air. When the air valve is opened, there is a hissing sound as the air escapes. However, after some minutes he realised that the hissing sound was not air escaping from the pipe but air rushing in. The pipe was in fact above the hydraulic gradient and was working as a siphon at that point and the valve was only there to let air out during the priming process. The pressure inside the pipe was in fact negative and so when he opened the valve, air was sucked and this de-primed the siphon. Realising his mistake, he quickly closed the valve and went on down to the farmhouse. The farmer was most upset. What a coincidence, his water supply had suddenly stopped, and an engineer just happened to be on hand to fix it for him!

4.7.3 Some practical siphons

If your car ever runs out of petrol, a siphon can be a useful means of taking some fuel from a neighbour's tank. Insert a flexible small diameter plastic tube into the tank and suck out all the air (making sure not to get a mouthful of petrol). When the petrol begins to flow, catch it in a container and then transfer it to your car. Make sure that the outlet is lower than the liquid level in the tank otherwise the siphon will not work.

Another very practical use for siphons is to detect leakage in domestic water mains (sometimes called rising mains) from the supply outside in the street into your home (Figure 4.5b). This can be important for those on a water meter who pay high prices. A leaky pipe in this situation would be very costly. The main valve to the house must first be closed. Then seal the cold water tap inside the house by immersing the outlet in a pan of water and opening the tap. If there is any leakage in the main pipe then the water will be siphoned back out of the pan into the main. The rate of flow will indicate the extent of the leakage.

4.7.4 Siphon limits

Siphons can be very useful in situations where the land topography is undulating between a reservoir and the water users. It is always preferable to locate a pipe below the hydraulic gradient by putting it in a deep trench but this may not always be practicable. In situations where siphoning is unavoidable, the pipeline must not be more than 7 m above the hydraulic grade line. Remember atmospheric pressure drives a siphon and the absolute limit is 10 m head of water. So, 7 m is a safe practical limit. When pipelines are located in mountainous regions, the limit needs to be lower than this due to the reduced atmospheric pressure.

The pressure inside a working siphon is less than atmospheric pressure and hence it is negative when referred to as a gauge pressure (measured above or below atmospheric pressure as the datum), for example, – 7 m head. Sometimes siphon pressures are quoted as absolute pressures (measured above vacuum pressure as the datum). So, –7 m gauge pressure is the same as +3 m absolute pressure. This is calculated as follows

gauge pressure = –7 m head

absolute pressure = atmospheric pressure + gauge pressure

= 10 – 7 = 3 m head absolute.

4.8 SELECTING PIPE SIZES IN PRACTICE

The development of λ as a pipe roughness coefficient is an interesting story and this nicely leads into the use of the Darcy–Weisbach formula for linking energy loss with the various pipe parameters. There are several examples using this formula in the boxes and they demonstrate well the effects of pipe length, diameter and velocity on energy loss. So, they are a useful learning tool.

Engineers in different industries and in different countries use other formulae as well as Darcy–Weisbach. Some are developed empirically to fit their particular circumstances. But a new generation of engineers are now replacing these with the Colebrook–White formula which is more fundamentally based and covers most commercially available pipes.

For completeness here is the Colebrook White formula for calculating λ

$$\frac{1}{\sqrt{\lambda}} = 2\log\left[\frac{k}{3.7d} + \frac{2.51}{R_e\sqrt{\lambda}}\right],$$

where k is the height of roughness elements in the pipe, d is the pipe diameter, and R_e is Reynolds number. Note that the value λ is on both sides of this equation and so an iterative solution is needed. Once λ is known then we use the Darcy–Weisbach formula to calculate discharge. Using the formula, is rather complicated, though modern spreadsheet design methods are making it simpler to use. An alternative is to use a design chart (Figure 4.8). These are easier to use and a range of pipe solutions can be assessed relatively quickly. Boxes 4.2 and 4.3 provide examples using the Darcy–Weisbach formula and design charts based on Colebrook–White formula.

BOX 4.2   EXAMPLE: CALCULATING PIPE DIAMETER USING DARCY–WEISBACH FORMULA

A 2.5 km long pipeline connects a reservoir to a smaller storage tank outside a town which then supplies water to individual houses. Determine the pipe diameter when the steady discharge required between the reservoir and the tank is 0.35 m³/s and the difference in their water levels is 30 m. Assume the value of λ is 0.03 (Figure 4.6).

This problem can be solved using the energy equation. The first step is to write down the equation for two points in the system. Point I is at the water surface of the main reservoir and point 2 is at the surface of the tank. Friction losses are important in this example and so these must also be included

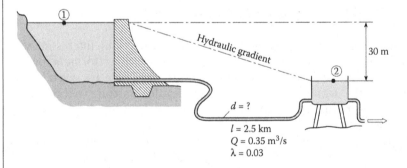

Figure 4.6 Calculating the pipe diameter.

$$\frac{p_1}{\rho g} + \frac{v_1^2}{2g} + z_1 = \frac{p_2}{\rho g} + \frac{v_2^2}{2g} + z_2 + h_f.$$

This equation can be greatly simplified. p_1 and p_2 are both at atmospheric pressure and are zero. The water velocities v_1 and v_2 in the two tanks are small and hence the kinetic energy terms are also small and can be ignored. This leaves just the potential energy terms z_1 and z_2 and the energy loss term h_f, so the equation simplifies to

$$h_f = z_1 - z_2.$$

Using the Darcy–Weisbach formula for h_f

$$h_f = \frac{\lambda l v^2}{2gd}.$$

Rearranging this equation

$$\frac{\lambda l v^2}{2gd} = z_1 - z_2.$$

Diameter d is unknown but so is the velocity in the pipe. So, first calculate velocity v using the continuity equation

$$Q = va.$$

Rearranging equation for v

$$v = \frac{Q}{a}.$$

Calculate area a

$$a = \frac{\pi d^2}{4},$$

and use this value to calculate v

$$v = \frac{4Q}{\pi d^2} = \frac{4 \times 0.35}{3.14 \times d^2} = \frac{0.446}{d^2}.$$

We need v^2 for the next step

$$v^2 = \frac{0.198}{d^4}.$$

Note that d is not known and so it is not yet possible to calculate a value for v. This can remain as an algebraic expression.

Put all the known information into the Darcy–Weisbach equation

$$\frac{\lambda l v^2}{2gd} = z_1 - z_2$$

$$\frac{0.03 \times 2{,}500 \times 0.198}{2 \times 9.81 \times d \times d^4} = 30.$$

Rearrange this to calculate d

$$d^5 = \frac{0.03 \times 2{,}500 \times 0.198}{2 \times 9.81 \times 30} = 0.0253.$$

Calculate the fifth root of 0.0253 to find d

$$d = 0.48 \text{ m} = 480 \text{ mm}.$$

The nearest pipe size to this would be 500 mm. So, this is the size of pipe needed to carry this flow between the reservoir and the tank.

This may seem rather involved mathematically but another approach, and perhaps a simpler one, is to guess the size of pipe and then put this into the equation and see if it gives the right value of discharge. This 'trial and error' approach is the way most engineers approach the problem. The outcome will show if the chosen size is too small or too large. A second or third guess will usually produce the right answer. If you are designing pipes on a regular basis you soon learn to 'guess' the right size for a particular installation. The design then becomes one of checking that your guess was the right one.

Try this design example again using the design chart in Figure 4.8 to see if you get the same answer.

BOX 4.3   EXAMPLE: CALCULATING DISCHARGE FROM A PIPELINE

A 200 mm diameter pipeline, 2,000 m long is connected to a reservoir and its outlet is 15 m below the reservoir water level and discharges freely into the atmosphere. Calculate the discharge from the pipe when the friction factor λ is 0.014 (Figure 4.7).

To solve this problem, use the energy equation between point 1 at the surface of the reservoir and point 2 just inside the water jet emerging from the pipe outlet

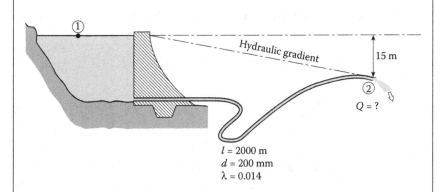

Figure 4.7 Calculating discharge from pipeline.

$$\frac{p_1}{\rho g} + \frac{v_1^2}{2g} + z_1 = \frac{p_2}{\rho g} + \frac{v_2^2}{2g} + z_2 + h_f.$$

This equation can be greatly simplified because p_1 is at atmospheric pressure and is zero. Also, p_2 is very near atmospheric pressure because the position of 2 is in the jet as it emerges from the pipe into the atmosphere. If it was above atmospheric pressure, the jet would flow laterally under the pressure. It does not do this and so the pressure can be assumed to be close to the atmospheric pressure. Therefore, p_2 is zero. The water velocity v_1 is zero in the reservoir and v_2 at the outlet is very small in comparison with the potential energy of 15 m, and hence this can also be assumed to be zero. This leaves just the potential energy terms z_1 and z_2 and the energy loss term h_f so the equation simplifies to

$$z_1 - z_2 = h_f.$$

So

$$z_1 - z_2 = \frac{\lambda l v^2}{2gd}.$$

Put in the known values and calculate velocity v

$$15 = \frac{0.014 \times 2{,}000 \times v^2}{2 \times 9.81 \times 0.2}$$

$$v^2 = \frac{15 \times 2 \times 9.81 \times 0.2}{0.014 \times 2{,}000} = 2.1$$

$$v = 1.45 \text{ m/s}.$$

Use the continuity equation to calculate the discharge

$Q = va$.

Calculate area a

$$a = \frac{\pi d^2}{4} = \frac{\pi 0.2^2}{4} = 0.031\,m^2.$$

Now calculate the discharge

$Q = 1.45 \times 0.031 = 0.045\,m^3/s$ or   45 L/s

4.8.1 Using hydraulic design charts

Pipe charts are now an increasingly common way of designing pipes. An excellent and widely used source is Hydraulic Design of Channels and Pipes (Hydraulics Research, 1990). This is a book of design charts based on the Colebrook–White equation. The equation best describes the transitional flow between smooth and rough pipe flow referred to in Section 4.4.1 and covers all commercially available pipes. An example of one of these charts is shown in Figure 4.8. It does not use λ values but expresses friction as the height of the roughness on the inside of a pipe. This chart is for a surface roughness of $k = 0.03$ mm and is representative of PVC pipes in reasonably good condition. The chart's range of flows is considerable; from less than 0.1 to 20,000 L/s (or 20 m³/s) with pipe diameters from 0.025 to 2.5 m. This should satisfy most pipe designers. In Box 4.4 is an example showing how to use the design chart.

There are four important practical points to note from the examples.

The first point refers to the first worked example which showed how mathematically cumbersome it can be to determine the diameter by calculation. The easier way is to do what most engineers do; they guess the diameter and then check by calculation that their chosen pipe is the right one. This might seem a strange way of approaching a problem but it is quite common in engineering. An experienced engineer usually knows what answer to expect, the calculation is just a way of confirming this. This is one of the basic unwritten laws of engineering – that you need to know the answer to the problem before you begin so that you know that you have the right answer when you get there. Real problems are not like those in the textbook which come with answers. Knowing you have the right answer comes largely from experience of similar design problems. New designers are unlikely to have this experience but they have to start somewhere and one way is to rely initially on the experience of others and to learn from them. This is the apprenticeship that all engineers go through to gain experience and become competent designers.

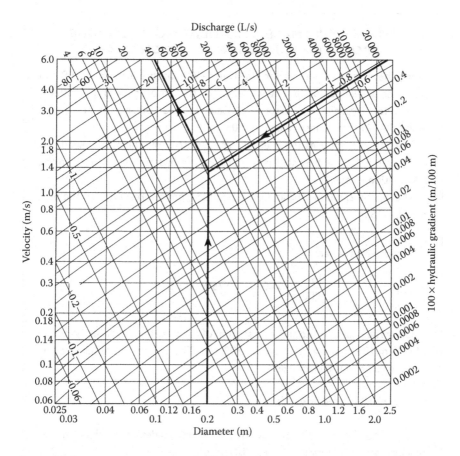

Figure 4.8 Typical pipe design chart for Colebrook White equation for k = 0.03.

The second point to note is that there is no unique pipe diameter that must be used in any given situation. If, for example, calculations show that a 100 mm pipe is sufficient then any pipe larger than this will also carry the flow. The question of which one to choose may be determined by other design criteria. One common guideline is to limit velocity to 1.6 m/s to avoid excessive head losses and limit water hammer problems. Another might be a limit on the head loss from friction. A third could be a limit on the pipe sizes available. There are only certain standard sizes which are manufactured and not all these may be readily available in some countries. A final deciding factor is cost – which pipe is the cheapest to buy and to operate?

The third point to consider is the value for pipe roughness. It is easy to choose the value for a new pipe but how long will it be before the roughness increases? What will the value be in say 10 years from now when the pipe is still being used? The roughness will undoubtedly increase through

BOX 4.4   EXAMPLE: CALCULATING DISCHARGE FROM A PIPELINE USING A DESIGN CHART

Using the same example as for the Darcy–Weisbach equation. A 200 mm diameter pipeline 2,000 m long is connected to a reservoir and the outlet is 15 m below the reservoir water level. Calculate the discharge from the pipe when the pipe roughness value k is 0.03 mm.

The design chart uses the hydraulic gradient to show the rate of head loss in a pipeline. So, the first step is to calculate the hydraulic gradient from the information given above.

The pipe is 2,000 m long and the head loss from the reservoir to the pipe outlet is 15 m, so

$$\text{hydraulic gradient} = \frac{h}{l}$$

$$\frac{15}{2,000} = 0.0075 = 0.75\,\text{m}/100\,\text{m}.$$

Using the chart, locate the intersection of the lines for a hydraulic gradient of 0.75 m/100 m and a diameter of 200 mm. This locates the discharge line and the value of the discharge

$$Q = 0.045\,\text{m}^3/\text{s or }45\,\text{L/s}.$$

Note that the chart can also be used in reverse to determine the diameter of a pipe and head loss for a given discharge.

general wear and tear. If lime scale deposits or algae slime build up on the inside of the pipe or the pipe is misused and damaged, the roughness will be significantly greater. So when choosing the most appropriate value for design, it is important to think ahead and what the roughness might be later in the life of the pipe. This is where engineering becomes an art and all the engineer's experience is brought to bear in selecting the right roughness value for design purposes. If you select a low roughness value, the pipe may not give good, long service. If you choose a high value, this will result in unnecessary expense of having pipes which are too big for the job in hand.

The final point, and often the most important to consider is cost. Small diameter pipes are usually cheaper than larger diameter pipes, but they require more energy to deliver the discharge because of the greater friction. This is particularly important when water is pumped and energy costs are high. The trade-off between the two must take account of both capital and operating costs if a realistic comparison is to be made between alternatives. This aspect of pipeline design is covered in more detail in Chapter 7 Pumps.

4.8.2 Sizing pipes for future demand

Pipes sizes are often selected using a discharge based on present water demands and little thought is given to how this might change in the future. Also, there is always a temptation to select small pipes to satisfy current demand simply because they are cheaper than larger ones. These two factors can lead to trouble in the future. If demand increases and higher discharges are required from the same pipe, the energy losses can rise sharply and so a lot more energy is needed to run the system.

As an example, a 200 mm diameter pumped pipeline, 500 m long, supplies a small town with a discharge of 50 L/s. Several years later the demand doubles to 100 L/s. This increases the velocity in the pipe from 1.67 to 3.34 m/s (i.e. it doubles) which pushes up the energy loss from 5 to 20 m (i.e. a fourfold increase). This increase in head loss plus the extra flow means that eight times more energy is needed to operate the system and extra pumps will be required. A little extra thought at the planning stage and a little more investment at the beginning could save a lot of extra pumping cost later.

Increasing the energy available is one way of increasing the discharge in a pipeline to meet future demand. But another way is to increase the effective diameter of the pipe. A practical way of doing this is to lay a second pipe parallel to the first one. It may not be necessary to lay the second pipe along the entire length. Pipes are expensive, and so from a cost point of view only the minimum length of parallel pipe should be laid to meet the demand (Figure 4.9). The discharge in pipe 1 will equal the discharge in pipe 2. This is the original pipeline carrying the (inadequate) discharge between the tanks – note the energy line which represents the uniform energy loss along the pipeline. Pipe 3 is the new pipe laid parallel to the original pipe and so the combined effect of pipes 2 and 3 is to increase the cross-sectional

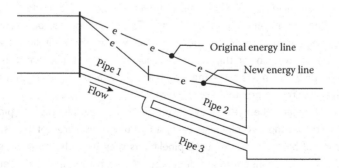

Figure 4.9 Parallel pipes can increase discharge with the same energy.

area carrying the discharge and decrease the energy loss along the parallel section of the pipeline (the velocity is lower because of the increased area). The effect of reducing the energy loss in the parallel section is to make more energy available to move water through pipeline 1 and so the overall discharge is increased. Note how the energy line for pipe 1 in the new system is steeper showing that it is carrying a higher discharge. The energy line for the parallel pipes has a gentler gradient due to the overall reduction in velocity in this section. The length of parallel pipe depends on the required increase in discharge. Should the discharge demand increase further in the future, the length of parallel pipeline can be extended to suit. The job for the designer is to decide on the diameter and length of pipe 3 (see Box 4.5).

BOX 4.5   EXAMPLE: CALCULATING LENGTH OF A PARALLEL PIPE

A 1,000 m long pipeline, 150 mm diameter supplies water from a reservoir to an offtake point. Calculate the discharge at the offtake when the head available is 10 m. Since the pipeline was installed, the water demand has doubled and so a parallel 250 mm diameter pipeline is to be installed alongside the original pipeline (Figure 4.9). Calculate the length of new pipe required to double the discharge. Assume the friction factor for the pipelines is $k = 0.03$ mm. Use the pipe design chart in Figure 4.8.

First calculate the original discharge. Calculate the hydraulic gradient and together with the pipe diameter determine the discharge from the pipe design chart. The data and the results are tabulated as follow(L/s)

Pipe	Pipe dia (mm)	Length (m)	Friction k (mm)	Hyd grad (m/100 m)	Discharge (L/s)
Original pipe	150	1,000	0.03	1.0	23

The demand has now doubled to 46 L/s. So, the additional 23 L/s is supplied by introducing a pipe of 250 mm diameter – pipe 3 – alongside the original pipeline but as yet of unknown length. This length cannot be calculated directly and requires some iteration. In other words, some intelligent guess work. It is convenient at this stage to divide the original pipeline into two parts – pipe 2 which has the same length as pipe 3, and pipe 1 which is from the reservoir to the point where the two parallel pipes join.

First determine the hydraulic gradient in the two parallel pipes – pipes 2 and 3 – so that the two pipelines carry a combined discharge of 46 L/s. The gradient will be the same for each pipeline as they have the same pressure at the points of connection and discharge. Mark on the pipe chart vertical lines representing the two pipe diameters. Look for a hydraulic gradient that intersects the pipe 'lines' so that the sum of the two discharges is 46 L/s. This occurs at a hydraulic gradient of 0.18 which results in discharges of 10 and 35 L/s totalling 45 L/s. This is close enough to 46 L/s.

Next, determine the hydraulic gradient for pipe 1 for a discharge of 46 L/s. From the chart this is 3.2.

Using this information, it is now possible to set up an equation to calculate the length of pipe 3. The sum of the head loss in pipes 1 and 3 (remember the loss in pipe 3 will be the same as pipe 2) is 10 m. So

$$h_1 - h_3 = 10 \text{ m.}$$

Now calculate h_1 and h_3

$$\frac{100\, h_1}{L_1} = \text{hydraulic gradient} = 3.2,$$

and so

$$h_1 = \frac{L_1 \times 3.2}{100}.$$

Similarly

$$h_3 = \frac{L_3 \times 0.18}{100}.$$

So

$$\frac{L_1 \times 3.2}{100} + \frac{L_3 \times 0.18}{100} = 10.$$

Both L_1 and L_3 are unknown. So to 'eliminate' one of the unknowns, substitute for L_1 in terms of L_3

$$L_1 = 1{,}000 - L_3.$$

Substitute for L_1 in the above equation

$$\frac{(1{,}000 - L_3) \times 3.2}{100} + \frac{L_3 \times 0.18}{100} = 10$$

$$(1{,}000 - L_3)3.2 + 0.18\, L_3 = 1{,}000$$

$$3{,}200 - 3.2\, L_3 + 0.18\, L_3 = 1{,}000$$

$$3.02\, L_3 = 2{,}200 \quad L_3 = \frac{2{,}200}{3.02} = 728 \text{ m.}$$

So the length of pipe 3 is 728 m. This is also the length of pipe 2. Pipe 1 will be 274 m. The table summarises the results:

Pipe	Pipe dia (mm)	Friction k (mm)	Hyd grad (m/100 m)	Discharge (L/s)	Length (m)
Pipe 1	150	0.03	3.20	46	274
Pipe 2	150	0.03	0.18	10	728
Pipe 3	250	0.03	0.18	35	728

This is one example where using a formula makes the problem easier to solve as it avoids the iterative approach. Try to solve the problem using the Darcy–Weisbach formula and the continuity equation with a λ value of 0.04 for the pipes.

4.9 PIPE NETWORKS

Most water supply systems are not just single pipes but comprise a network of pipes. These supply water to several dwellings in a village or town (Figure 4.10). Some networks are simple and involve just a few pipes but some are quite complicated involving many different pipes and connections. Sometimes, the pipes form a ring or loop and this ensures that if one section of pipe fails for some reason then flow can be maintained from another direction. It also has hydraulic advantages. Each offtake point is supplied from two directions and hence the pipe sizes in the ring can be smaller if the point was fed from a single pipe.

The simplest example of a ring or loop network is a triangular pipe layout (Figure 4.11). Water flows into the loop at point A and flows out at points B and C. So the water flows away from point A towards B and C. But

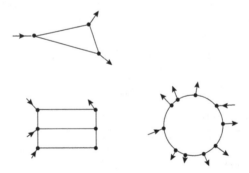

Figure 4.10 Pipe networks.

there will also be a flow in pipe BC and this could be in the direction BC or CB depending on the pressure difference between B and C. If the pressure at B is higher than C then there will be a flow from B to C. Conversely, if the pressure at C is higher than the pressure at B then the flow will be from C to B.

There are three rules to solving the problem of pressures and discharges in a network:

1. The sum of all the discharges at a junction is zero
2. The flow in one leg of a network will be in the direction of the pressure drop
3. The sum of the head losses in a closed loop will be zero for a start and finish at any junction in the loop.

These rules apply to all networks and not just the simple ones. However, the calculations can get rather involved. An example of how the rules are applied to a simple network is shown in Box 4.6.

BOX 4.6   EXAMPLE: CALCULATING DISCHARGES AND PRESSURES IN A PIPE NETWORK

A simple network of three pipes forms a triangle ABC. The following data are available (Figure 4.11):

Pipe	Diameter (mm)	Length (m)	Friction factor k (mm)
AB	300	2,000	0.03
BC	150	1,200	0.03
CA	450	2,050	0.03

The discharge entering the system at point A is 100 L/s and this meets the demand at B of 50 L/s and at C of 50 L/s.

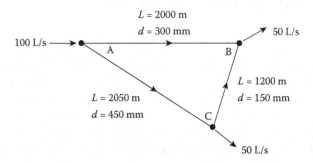

Figure 4.11 Triangular pipe network.

Calculate the discharges in each pipe and the pressures at B and C if A is supplied at a pressure of 100 m head of water. Use the pipe design chart in Figure 4.8.

Start by looking at the data, apply some common sense and rule (1) to get an assessment of the likely discharges in each pipe. AB and AC are approximately the same in length and AC has a large diameter. So assume Q_{AC} is 60 L/s, Q_{AB} is 40 L/s and Q_{CB} is 10 L/s. Note the assumed flow directions indicated by the arrows.

Next, calculate the head loss in each pipe using the pipe chart in Figure 4.8. Notice how the discharges in pipes BC and CA are listed as negative. The sign comes from considering the discharges positive in a clockwise direction around the loop.

Pipe	Diameter (mm)	Discharge (L/s)	Hydraulic gradient (m per 100 m)	Length (m)	Head loss (m)
AB	300	40	0.14	2,000	+2.80
BC	150	−10	0.2	1,200	−2.40
CA	450	−60	0.025	2,050	−0.51

Applying rule (3) from the start point A and moving in a clockwise direction add up all the head losses in the pipes, that is, $2.80 - 2.40 - 0.51 = -0.11$ m. This sum should come to zero.

To make the above sum come to zero, slightly decrease the discharge in pipe AB, recalculate the head losses again and see if the sum of the head losses comes to zero. In this case, the value is close to zero so this suggests that the discharge values chosen are close to the right ones. So, there is no need for further iteration.

4.10 MEASURING DISCHARGE IN PIPES

Discharges in pipelines can be measured using a *venturi meter* or an *orifice plate* (Figure 4.12). Both devices rely on changing the components of the total energy of flow from which discharge can be calculated (see Section 3.7). The venturi meter was developed by an American, Clemens Herschel (1842–1930) who was looking for a way to measure water abstraction from a river by industrialists. Although the principles of this measuring device were well established by Bernoulli, it was Herschel who, being troubled by unlicensed and unmeasured abstractions by paper mills, developed it into the device we use today.

A venturi meter comprises a short, narrow section of pipe (throat) followed by a gradually expanding tube. This causes the flow velocity to

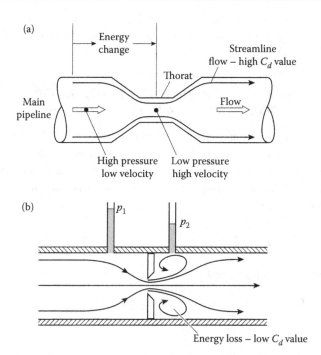

Figure 4.12 Measuring discharge in pipelines. (a) Venturi meter. (b) Orifice meter.

increase (remember continuity) and hence the kinetic energy also increases. As the total energy remains the same throughout the system, it follows that there must be a corresponding reduction in pressure energy. By measuring this change in pressure using a pressure gauge or a manometer and using the continuity and energy equations, the following formula for discharge in the pipe can be obtained

$$Q = C_d a_1 \sqrt{\frac{2gH}{m^2 - 1}}$$

$$m = \frac{a_1}{a_2},$$

where a_1 is the area of main pipe (m²), a_2 is the area of venturi throat (m²), H is the head difference between pipe and throat (m), g is the gravity constant (9.81 m/s²) and C_d is the coefficient of discharge.

This is the theory but in practice there are some minor energy losses in a venturi and so a coefficient discharge C_d is introduced to obtain the true discharge. Care is needed when using this formula. Some textbooks quote the formula in terms of a_2 rather than a_1 and this changes several of the terms. It is the same formula from the same fundamental base but it can be confusing. The safest way is to avoid the formula and work directly from the energy and continuity equations. A derivation of the formula and an example of calculating discharge working from energy and continuity are shown in Boxes 4.7 and 4.8.

BOX 4.7   DERIVATION: FORMULA FOR DISCHARGE IN A VENTURI METER

First, write down the energy equation for the venturi meter. Point 1 is in the main pipe and point 2 is located in the throat of the venturi (Figure 4.13). It is assumed that there is no energy loss between the two points. This is a reasonable assumption as contracting flows suppress turbulence which is the main cause of energy loss.

$$\frac{p_1}{\rho g} + \frac{v_1^2}{2g} + z_1 = \frac{p_2}{\rho g} + \frac{v_2^2}{2g} + z_2.$$

As the venturi is horizontal

$$z_1 = z_2,$$

and so

$$\frac{p_1}{\rho g} + \frac{v_1^2}{2g} = \frac{p_2}{\rho g} + \frac{v_2^2}{2g}.$$

Now, rearrange this equation so that all the pressure terms and all the velocity terms are brought together

$$\frac{p_1}{\rho g} - \frac{p_2}{\rho g} = \frac{v_2^2}{2g} - \frac{v_1^2}{2g}.$$

The left-hand side of this equation is the pressure difference between points 1 and 2 which can be measured using pressure gauges or a differential

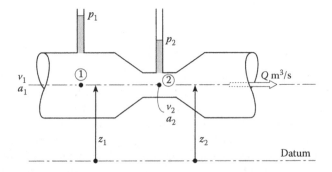

Figure 4.13 Venturi meter for measuring discharge.

manometer. This is a manometer with one limb connected to the pipe and the other limb connected to the throat (see Section 2.9).

At this point it is not possible to calculate the velocities because both v_1 and v_2 are unknown. A second equation is needed to do this – the continuity equation.

Write the continuity equation for points 1 and 2 in the venturi

$$a_1 v_1 = a_2 v_2.$$

Rearrange this

$$v_2 = \frac{a_1}{a_2} v_1.$$

Now a_1 and a_2 are the cross-sectional areas of the pipe and venturi, respectively, and can be calculated from the pipe and venturi throat diameters, respectively. Substituting for v_1 in the energy equation

$$\frac{p_1}{\rho g} - \frac{p_2}{\rho g} = \left(\frac{a_1^2}{a_2^2}\right)\frac{v_1^2}{2g} - \frac{v_1^2}{2g}$$

$$\frac{p_1}{\rho g} - \frac{p_2}{\rho g} = \frac{v_1^2}{2g}\left(\frac{a_1^2 - a_2^2}{a_2^2}\right).$$

Put

$$H = \frac{p_1}{\rho g} - \frac{p_2}{\rho g},$$

where H is the difference in head between the pipe (point 1) and the venturi throat (point 2)

Rearrange the equation for v_1

$$v_1 = \sqrt{2gH}\left(\frac{a_2}{\sqrt{a_1^2 - a_2^2}}\right).$$

Use the continuity equation to calculate discharge

$$Q = a_1 v_1,$$

and so

$$Q = \sqrt{2gH}\left(\frac{a_1 a_2}{\sqrt{a_1^2 - a_2^2}}\right).$$

Put

$$m = \frac{a_1}{a_2},$$

and so

$$Q = a_1 \sqrt{\frac{2gH}{m^2 - 1}}.$$

Introduce a coefficient of discharge C_d

$$Q = C_d a_1 \sqrt{\frac{2gH}{m^2 - 1}}.$$

In the case of the venturi meter $C_d = 0.97$, which means that the energy losses are small and the energy theory works very well ($C_d = 1.0$ would mean the theory was perfect). For orifice plates, the same theory and formula can be used but the value of C_d is quite different at $C_d = 0.6$. The theory is not so good for this case because there is a lot of energy loss (Figure 4.12b). The water is not channelled smoothly from one section to another as in the venturi but is forced to make abrupt changes as it passes through the orifice and expands downstream. Such abruptness causes a lot of turbulence which results in energy loss. (The C_d value is similar to that for orifice flow from a tank – see Section 3.6).

BOX 4.8 EXAMPLE: CALCULATING DISCHARGE USING A VENTURI METER

A 120 mm diameter venturi meter is installed in a 250 mm diameter pipeline to measure discharge. Calculate the discharge when the pressure difference between the pipe and the venturi throat is 2.5 m of head of water and C_d is 0.97.

Although, there is a formula for discharge, it can be helpful to work from first principles. Not only does this reinforce the principle but it also avoids possible errors in using a rather involved formula which can easily be misquoted. So, this example is worked from the energy and continuity equations.

First step is to write down the energy equation

$$\frac{p_1}{\rho g} + \frac{v_1^2}{2g} + z_1 = \frac{p_2}{\rho g} + \frac{v_2^2}{2g} + z_2.$$

As the venturi is horizontal

$$z_1 = z_2,$$

and so

$$\frac{p_1}{\rho g} + \frac{v_1^2}{2g} = \frac{p_2}{\rho g} + \frac{v_2^2}{2g}.$$

Now rearrange this equation so that all the pressure terms and all the velocity terms are brought together

$$\frac{p_1}{\rho g} - \frac{p_2}{\rho g} = \frac{v_2^2}{2g} - \frac{v_1^2}{2g}.$$

But

$$\frac{p_1}{\rho g} - \frac{p_2}{\rho g} = 2.5 \text{ m}.$$

Remember that the pressure terms are in m head of water and it is the difference that is important and not the individual pressures.
and so

$$\frac{v_2^2}{2g} - \frac{v_1^2}{2g} = 2.5 \text{ m}.$$

It is not possible to solve this equation directly as both v_1 and v_2 are unknown. So, use continuity to obtain another equation for v_1 and v_2

$$a_1 v_1 = a_2 v_2.$$

Rearrange this

$$v_2 = \frac{a_1}{a_2} v_1.$$

Next step is to calculate the areas a_1 and a_2
Area of pipe

$$a_1 = \frac{\pi d_1^2}{4} = \frac{\pi 0.25^2}{4} = 0.05 \text{ m}^2.$$

Area of venturi

$$a_2 = \frac{\pi d_2^2}{4} = \frac{\pi 0.12^2}{4} = 0.011 \, m^2$$

$$v_2 = \frac{0.05}{0.011} v_1 = 4.55 \, v_1.$$

Substitute this value for v_2 in the energy equation

$$\frac{4.55^2 v_1^2}{2g} - \frac{v_1^2}{2g} = 2.5 \, m$$

$$20.7 \, v_1^2 - v_1^2 = 2.5 \times 2 \times 9.81 = 49.05$$

$$v_1 = \sqrt{\frac{49.05}{19.7}} = 1.57 \, m/s.$$

Calculate Q

$$Q = C_d v_1 a_1$$
$$= 0.97 \times 1.57 \times 0.05 = 0.076 \, m^3/s.$$

4.11 MOMENTUM IN PIPES

The momentum equation is used in pipe flow to calculate forces on pipe fittings such as nozzles, pipe bends and valves. In more advanced applications it is used in the design of pumps and turbines where water flow creates forces on pump and turbine impellers.

To solve force and momentum problems, a concept known as the *control volume* is used. This is a way of isolating part of a system being investigated so that the momentum equation can be applied to it. To see how this works, an example is given in Box 4.9 showing how the force on a pipe reducer (or nozzle) can be calculated.

BOX 4.9   EXAMPLE: CALCULATING THE FORCE ON A NOZZLE

A 100 mm diameter fire hose discharges 15 L/s from a 50 mm diameter nozzle. Calculate the force on the nozzle (Figure 4.14).

To solve this problem, all three hydraulic equations are needed; energy, continuity and momentum. The energy and continuity equations are needed

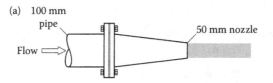

(b)

$$F_1 = p_1 a_1 \Longrightarrow \qquad \Longleftarrow F = \qquad \Longleftarrow F_2 = p_2 a_2$$

Control volume

Figure 4.14 Calculating the force on a nozzle. (a) Nozzle details. (b) Control volume.

to calculate the pressure in the 100 mm pipe and momentum is then used to calculate the force on the nozzle.

The first step is to calculate the pressure p_1 in the 100 mm pipe. Use the energy equation

$$\frac{p_1}{\rho g} + \frac{v_1^2}{2g} + z_1 = \frac{p_2}{\rho g} + \frac{v_2^2}{2g} + z_2.$$

The pressure in the jet as it emerges from the nozzle into the atmosphere is p_2. The jet is at the same pressure as the atmosphere and so the pressure p_2 is zero. The potential energies z_1 and z_2 are equal to each other because the nozzle and the pipe are horizontal and so they cancel out.

So, the energy equation becomes

$$\frac{p_1}{\rho g} = \frac{v_2^2}{2g} - \frac{v_1^2}{2g}.$$

The value of p_1 is unknown and so are v_1 and v_2. So, the next step is to calculate the velocities from the discharge equation

$$\frac{Q}{a_1}$$

$$v_1 = \frac{Q}{a_1} \text{ and } v_2 = \frac{Q}{a_2}.$$

Area of pipe

$$a_1 = \frac{\pi d_1^2}{4} = \frac{\pi 0.1^2}{4} = 0.0078 \text{ m}^2.$$

Area of jet

$$a_2 = \frac{\pi d_2{}^2}{4} = \frac{\pi 0.05^2}{4} = 0.0019 \text{ m}^2.$$

Next calculate the velocities

$$v_1 = \frac{Q}{a_1} = \frac{0.015}{0.0078} = 1.92 \text{ m/s}.$$

And

$$v_2 = \frac{Q}{a_1} = \frac{0.015}{0.0019} = 7.9 \text{ m/s}.$$

Put all the known values into the energy equation

$$\frac{p_1}{\rho g} = \frac{7.9^2}{2 \times 9.81} - \frac{1.92^2}{2 \times 9.81} = 2.99 \text{ m}$$

$$p_1 = 2.99 \times 1,000 \times 9.81 = 29,333 \text{ N/m}^2.$$

The final step is to calculate the force F on the nozzle using the momentum equation. To do this, the concept of the *control volume* isolates that part of the system being investigated. All the forces which help to maintain the control volume are then identified. F_1 and F_2 are forces due to the water pressure in the pipe and the jet, and F is the force on the water on the reducer. Although the force F_2 is shown acting against the flow, remember that there is an equal and opposite force acting in the direction of the flow (Newton's third law).

Use the momentum equation

$$F_1 - F_2 - F = \rho Q(v_2 - v_1).$$

F is the force of water on the nozzle but at this stage both F_1 and F_2 are also unknown. So, the next step is to calculate F_1 and F_2 using known values of the pressure and the pipe area

$$F_1 = p_1 \times a_1$$
$$= 29,332 \times 0.0078 = 228.8 \text{ N},$$

and

$$F_2 = p_2 \times a_2.$$

But p_2 is 0 and so

$$F_2 = 0.$$

Finally, all the information for the momentum equation is available to calculate F

$$228.8 - F - 0 = 1,000 \times 0.015 \ (7.9 - 1.92)$$
$$F = 228.8 - 89.7$$
$$F = 139 \, N.$$

As this is a fire hose, a firm grip would be required to hold it in place. If a fireman let go of the nozzle, the unbalanced force of 139 N would cause it to rapidly shoot forwards and this could do serious injury if it hit someone. A larger nozzle and discharge would probably need two firemen to hold and control it.

4.11.1 Pipe bends

Forces, often large ones, also occur at pipe bends and this is not always appreciated. For example, a 90° bend on a 0.5 m diameter pipe operating at 30 m head carrying a discharge of 0.3 m³/s would produce a force of 86 kN. This is a large thrust (over 8 tons), which means that the bend must be held firmly in place if it is not to move and burst the pipe (Figure 4.15).

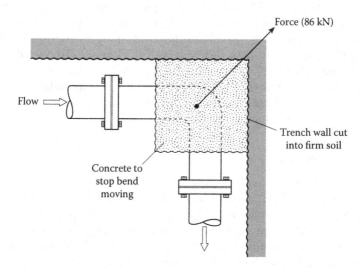

Figure 4.15 Force on a pipe bend.

A good way to deal with this is to bury the pipe below ground and encase it in concrete. The side of the trench must also be very firm for the concrete to push against.

4.12 PIPE MATERIALS

Pipes are a major cost in any water supply or irrigation scheme. They must carry the design flow, resist all external and internal forces, be durable and have a long useful life. So, it is important to know something about what kind of pipes are available, how they are installed and used, what valves and fittings are needed and how to test them once they are in place.

4.12.1 Specifying pipes

Pipe manufacture is normally controlled by specifications laid down by national and international standards organisations. In the United Kingdom, the British Standards Institute (BSI) has developed its own standards for pipes, although a notable exception to this is the aluminium pipe used in many irrigation systems. The standards were developed for the water supply industry that demand high-quality manufacture and rigorous testing of all pipes and fittings. The International Standards Organisation (ISO) does a similar job on an international scale. Many standards are the same and some BSI publications, for example, also have an ISO number.

Although the internal bore of a pipe determines its flow capacity, it is not generally used to specify pipe size. The reason for this is that pipes are also classified according to pressure which means they have different wall thicknesses. If the outside diameter of a pipe is fixed, which is often the case from a pipe joining point of view, the internal bore will be different for different wall thicknesses. To add to this confusion, different pipe materials (e.g. PVC, steel) will have different wall thicknesses. To overcome this problem, manufacturers' quote a *nominal pipe size*, which in many cases is neither the inside nor the outside diameter! It is just an indication of the pipe diameter for selection purposes. It is also normal to specify the safe working pressure at which the pipes can be used.

4.12.2 Materials

Pipes are made from a wide variety of materials but the most common in water systems are steel, asbestos cement and various plastics. Steel pipes tend to be used only where very high pressures are encountered or in conjunction with other pipe materials where extra strength is needed such as under roads or across ditches. Larger pipes are made from steel plate bent or rolled to shape and butt welded. Small sizes up to 450 mm are made from hot steel ingots which are pierced and rolled into a cylinder of the right dimensions.

Corrosion is a problem with steel and so pipes are wrapped with bituminous materials or galvanising. Buried pipes often use cathodic protection. Corrosion is an electrolytic process set up between the pipe and the surrounding soil. It is a bit like a battery with the steel pipe acting as an anode which gradually corrodes. The process is reversed by making the pipe a cathode either by connecting it to an expendable anodic material such as magnesium or by passing a small electric current through the pipe material.

Asbestos cement pipes are still used in some countries for underground mains though no longer installed in Europe because of health risks. They are more associated with manufacture rather than use. They are made from cement and asbestos fibre mixed in a slurry and deposited layer upon layer on a rolling mandrel. The pipe is then dipped in cold bitumen for protection and the end turned down to a specific diameter for jointing purposes. Unlike steel, asbestos cement does not corrode but it is easily damaged by shock loads and must be handled with care. Sometimes damage is not always apparent as in the case of hairline fractures. These only show up when the pipe is being site tested and leaks occur. Joints are made using an asbestos cement collar and rubber sealing rings similar to those used for PVC pipes. Bends and fittings cannot be made from asbestos cement and so ductile iron is used. An important criterion is that the outside diameter of the ductile steel must be the same as the asbestos cement so they can be effectively joined together.

Several plastic materials are in common use for making pipes. Unplasticised polyvinyl chloride (UPVC) is a rigid material and an alternative to asbestos cement, which is falling out of favour because of the health risks associated with asbestos fibres. Pipes come in various pressure classes and are colour coded for easy recognition. They are virtually corrosion free, light in weight, flexible and easily jointed by a spigot and socket system using either a chemical solvent or rubber ring to create a seal. But laying PVC pipes need care. They are easily damaged by sharp stones in the soil and distorted by poor compaction of material on the bottom and sides of trenches. Hydraulically, PVC has a very low friction characteristic but it can be easily damaged internally by sand and silt particles in the flowing water.

In contrast, polythene pipe is very flexible and comes in high-density and low-density forms. Low-density pipe is cheaper than high density and is used extensively for trickle and some sprinkler irrigation systems where pipes are laid out on the ground. The pipe wall is thin and so it is easily damaged by sharp tools or animals biting through it. However, it will stand up to quite high water pressures. High-density pipes are much stronger but more expensive, and are used extensively for domestic water supply systems.

4.13 PIPE FITTINGS

Pipelines require a wide range of valves and special fittings to make sure the discharge is always under control. Some of the more important are sluice valves, air valves, non-return valves and control valves (Figure 4.16).

Sluice valves are essentially on-off valves. They can be used to control pressure and discharge but they are rather crude and need constant attention. It is only the last 10% of the gate opening that has any real controlling influence. Moving the gate over the remaining 90% does not really influence the discharge, it just changes the energy from pressure to kinetic energy so the flow goes through the narrow opening faster. The valve body contains a gate which can be lowered using a screw device to close off the flow. The gate slides in a groove in the valve body and relies on the surfaces of the gate and the groove being forced together to make a seal. It was developed over 100 years ago for the water industry and has not changed materially since then. The gate can be tapered so that it does not jam in the gate guide but the taper gate must be opened fully otherwise it may start to vibrate once it is partially open. Sluice valves do require some effort to open them as water pressure builds up on one side producing an unbalanced head. The term 'cracking open' a valve is sometimes used to describe the initial effort needed to get the valve moving. A 100 mm valve fitted with a hand wheel is difficult to open when there is an unbalanced pressure of over 8 bar. Larger valves often used gearing mechanisms to move them, and so opening and closing valves can be a slow business. This is an advantage as it reduces the problems of water hammer (see Section 4.14).

Butterfly valves do the same job as sluice valves – they consist of a disc which rotates about a spindle across the diameter of the pipe. They are easier to operate than sluice valves requiring less force to open them. In spite of this, they have not always been favoured by designers because the disc is an obstacle to the flow where debris can accumulate. It is also possible to close a small butterfly valve quite rapidly and this can cause water hammer problems.

Air valves are a means of letting unwanted air out of pipelines. Water can contain large volumes of dissolved air which can come out of solution when the pressure drops, particularly at high spots on pipelines (See Section 4.7). If air is allowed to accumulate, it can block the flow or at least considerably reduce it. Air valves allow air to escape. There are two types depending on whether the pipeline is above or below the hydraulic gradient. For pipes below the hydraulic gradient where the water pressure is positive, the most common situation, a ball valve is used. When air collects in the pipe, the ball falls onto its seating allowing the air to escape. As the air escapes, water rises into the valve which pushes the ball up to close off the valve. When the pipeline is above the hydraulic gradient, the pressure inside the pipe is negative (below atmospheric pressure). In this case, the ball valve would not work as air would be sucked in and this would allow

(a)

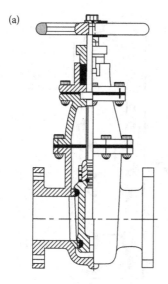

(b)

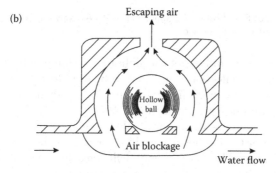

(c)

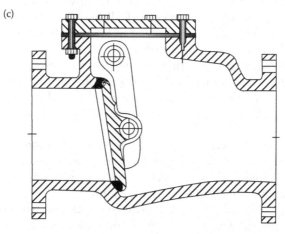

Figure 4.16 Pipe fittings. (a) Sluice valve. (b) Air valve. (c) Non-return valve. (d) Pressure control valve.   (*Continued*)

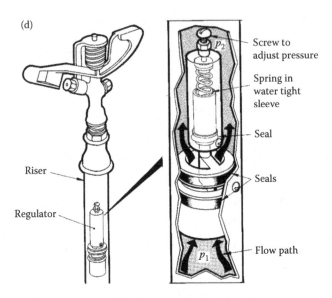

Figure 4.16 (Continued) Pipe fittings. (a) Sluice valve. (b) Air valve. (c) Non-return valve. (d) Pressure control valve.

the pipe to fill with air and break the siphon. A manual valve is needed to remove the air but a bellows fitting can also be used to suck out unwanted air. This process is very much like the priming described both for siphons (see Section 4.7) and pumps (see Section 7.4).

The non-return valve – sometimes called a reflux valve – does what it says. It allows water to flow one way only and prevents the return flow once the main flow stops. These are essential fittings on pump delivery pipes to prevent damage from water hammer, and on water supply pipelines to prevent contamination from being 'sucked' into the pipe when it is being shut down. An example is an outside tap for watering the garden with a hose pipe. If for some reason another tap on the same domestic supply system is opened at the same time, the flow in the hose pipe can stop and even be reversed. This can suck soil particles and harmful bacteria into the pipeline and contaminate the flow. To avoid this problem all outside taps should be fitted with a non-return valve – in the United Kingdom it is a legal requirement.

Control valves are available for a wide range of control issues such as discharge and pressure control, pressure reducing, pressure sustaining and surge control. Some valves are very sophisticated and expensive but each works on a simple principle. One simple example is a valve to regulate pressure under agricultural sprinklers. Sprinklers work best when the operating pressure is constant and at the value recommended by the manufacturer. A small regulator placed under each sprinkler ensures that each sprinkler operates at the desired pressure. Normally, pressures p_1 and p_2 are equal and close to the sprinkler operating pressure (Figure 4.16d). But if the pressure p_1 rises then p_2 also starts to rise. This pushes down the sleeve closing up the waterway and reducing the flow to the sprinkler. The effect of this is to reduce the pressure p_2. If p_1 starts to fall then p_2 also falls, the sleeve opens and allows more flow through to maintain the pressure p_2. The regulated pressure p_2 is controlled by a spring in compression inside the sleeve. This can be adjusted using a small screw. For example, if a higher pressure p_2 is required then the screw is turned clockwise increasing the spring compression. This stops the movement of the sleeve until the higher pressure p_2 is reached. Although this is a simple mechanism, most pressure regulating valves work on this principle. Note that pressure control valves will only reduce the pressure; they cannot increase it beyond the working pressure in the pipeline.

4.14 WATER HAMMER

Most people will already have experienced this but may not have realised it nor appreciated the seriousness of it. When a domestic water tap is turned off quickly, there is sometimes a loud banging noise in the pipes and sometimes the pipes start to vibrate. The noise is the result of a high-pressure wave which moves rapidly through the pipes as a result of the rapid closure of the tap. This is known as *water hammer*, and although it may not be too serious in domestic plumbing, it can have disastrous consequences in larger pipelines and it may result in pipe bursts.

Water hammer occurs when flowing water is suddenly stopped. It behaves in a similar way to traffic flowing along a road, when suddenly one car stops for no clear reason (Figure 4.17a). The car travelling close behind then crashes into it and the impact causes the cars to crumple. The next one crashes into the other two and so on until there is quite a pile up. Notice that all the cars do not crash at the same time. A few seconds pass between each impact and so it takes several seconds before they all join the pile up. If you are watching this from a distance, it would appear as if there was a wave moving up the line of cars as each joins the pile up. The speed of the wave is equal to the speed at which successive impacts occur. It is worth pointing out that cars are designed to collapse on impact so as to absorb the kinetic energy. If they were built more rigidly, then all the energy on impact would be transferred to the driver and the passengers and not even seat belts would hold you in such circumstances.

Now, imagine water flowing along a pipeline at the end of which is a valve that is closed suddenly (Figure 4.17b). If water was not compressible then it would behave like a long solid rod and would crash into the valve with such enormous force (momentum change) that it would probably destroy the valve. Fortunately, water is compressible and it behaves in a similar manner to the vehicles, it squashes on impact. Think of the flow

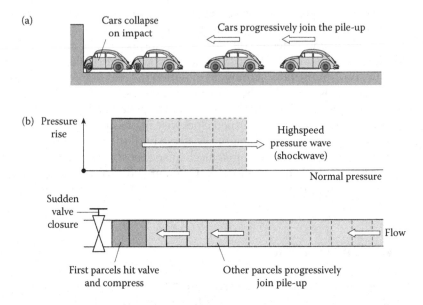

Figure 4.17 Water hammer. (a) Car crash analogy. (b) Pressure rise and shock-wave when valve is rapidly closed. (c) Sudden valve closure.

being made up of small 'parcels' of water. The first parcel hits the valve and compresses (like the first car); the second crashes into the first and compresses and so on until all the water is stopped (Figure 4.17b). This does not happen instantly but takes several seconds before all the water feels the impact and stops. The result is a sudden, large pressure rise at the valve and a pressure wave which travels rapidly along the pipe. This is referred to as a *shock wave* because of its suddenness.

The pressure wave is not just one way. Once it reaches the end of the pipeline, it reflects back towards the valve again. It is like a coiled spring that moves back and forth and gradually stops. This oscillating motion can go on for several minutes in a pipe until friction slowly reduces the pressure back to the normal operating level.

The extent of the pressure rise depends on how fast the water was travelling (velocity) and how quickly the valve was closed. It does not depend on the initial pipeline pressure as is often thought. It can be calculated using a formula developed by Nicholai Joukowsky (1847–1921) who carried out the first successful analysis of this problem

$$\Delta h = \frac{cv}{g},$$

where Δh is rise in pressure (m), c is velocity of the shock wave (m/s), v is water velocity (m/s) and g is gravity constant (9.81 m/s²).

The shock wave travels at very high velocity between 1,200 and 1,400 m/s. It depends on the diameter of the pipe and the material from which the pipe is made as some materials absorb the energy of compression of the water better than others. An example in Box 4.10 shows just how high the pressure can rise.

The example of course, is an extreme one as it is ifficult to close a valve instantaneously. It is also wrong to assume that the pipe is rigid. All materials stretch when they are under pressure and so the pipe itself will absorb some of the pressure energy by expanding. All these factors help to reduce the pressure rise but they do not stop it. Even if the pressure rise is only half the above value (say 6 bar), it is still a high pressure to suddenly cope with and this too is enough to burst the pipe. When pipes burst they usually open up along their length rather than around their circumference. They burst open rather like unzipping a coat.

Reducing water hammer problems is similar to reducing car crash problems. When cars are moving slowly then the force of impact is not as great. Also, when the first car slows down gradually then the others are unlikely to crash into it. Similarly, when water is moving more slowly the pressure

---

BOX 4.10 EXAMPLE: CALCULATING PRESSURE RISE IN A PIPELINE DUE TO WATER HAMMER

Using the Joukowsky equation, determine the pressure rise in a pipeline when it is suddenly closed. The normal pipeline velocity is 1.0 m/s and the shock wave velocity is 1,200 m/s. If the pipeline is 10 km long, determine how long it takes for the pressure wave to travel the length of the pipeline.

Using the Joukowsky equation

$$\text{Pressure rise } \Delta h = \frac{cv}{g}$$

$$= \frac{1,200 \times 1.0}{9.81} = 122 \text{ m.}$$

So, the pressure rise would be 122 m head of water or 12.2 bar. This is on top of the normal operating pressure and could well cause the pipe to burst.

Calculate the time it takes the wave to travel the length of the pipe

$$\text{time} = \frac{\text{distance}}{\text{velocity}} = \frac{10,000}{1,200} = 8 \text{ s.}$$

It takes only 8 s for the shock wave to travel the 10 km length of the pipe and 16 s for this to return to the valve.

rise when a valve closes is reduced (see Joukowsky equation). This is one of the reasons why most pipeline designers restrict velocities to below 1.6 m/s so as to reduce water hammer problems. Also, when valves are closed slowly, water slows down gradually and there is little or no pressure rise along the pipe.

In summary to reduce the effects of water hammer:

- Make sure water velocities are low (below 1.6 m/s)
- Close control valves slowly

In some situations it is not possible to avoid the sudden closure of a pipeline. For example, if a heavy vehicle drives over pipes laid out on the ground, such as might occur with fire hoses, it will squash them and stop the flow instantly. A similar situation can occur on farms where mobile irrigation machines use flexible pipes. It is easy for a tractor to accidentally drive over a pipe without realising that the resulting pressure rise can split open the pipeline and cause a lot of damage. In such situations where there are pipes on the ground and vehicles about, it is wise to use pipe bridges.

Some incidences though are not always easy to foresee. On a sprinkler irrigation scheme in East Africa, pipe bridges were used to allow tractors to cross pipelines. But an elephant got into the farm and walked through the crop. It trod on an aluminium irrigation pipe and squashed it flat resulting in an instantaneous closure. This caused a massive pressure rise upstream and several pipes burst open!

4.15 SURGE

Surge and water hammer are terms that are often confused because one is caused by the other. Surge is the large mass movement of water that sometimes takes place as a result of water hammer. It is much slower and can last for many minutes whereas water hammer may only last for a few seconds.

An example of the difference between the two can be most easily seen in a hydroelectric power station (Figure 4.18). Water flows down a pipeline from a large reservoir and is used to turn a turbine which is coupled to a generator that produces electricity. Turbines run at high speeds and require large quantities of water and so the velocities in the supply pipe can be very high. The demand for electricity can vary considerably over very short periods and problems occur when the demand falls and one or more of the turbines have to be shut down quickly. This is done by closing the valve on the supply pipe and this can cause water hammer. To protect a large part of the pipeline, a *surge tank* is located as close to the power station as possible. This is a vertical chamber many times larger than the pipeline diameter. Water no longer required for the turbine is diverted into the tank and any water hammer shock waves coming up from the valve closure are absorbed

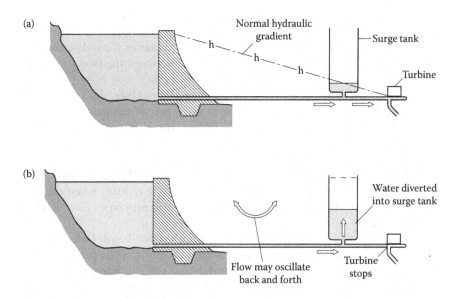

Figure 4.18 Surge in pipelines. (a) Normal operation. (b) Turbine shut down.

by the tank. Thus, water hammer is confined to the pipeline between the turbine and the tank and so only this length of pipe needs to be constructed to withstand the high water hammer pressures. Gradually, the tank fills with water and the flow from the reservoir slows down and eventually stops. Usually, the rushing water can cause the tank to overfill. In such cases water may flow back and forth between the tank and the reservoir for several hours. This slow but large movement of water is called *surge* and although it is the result of water hammer, it is quite different in character.

Surge can also cause problems in pumping mains and these are discussed more fully in Section 7.14.

4.16 SOME EXAMPLES TO TEST YOUR UNDERSTANDING

1. A 150 mm pipeline is 360 m long and has a friction factor $\lambda = 0.02$. Calculate the head loss in the pipeline using the Darcy–Weisbach formula when the discharge is 0.05 m³/s. Calculate the hydraulic gradient in m/100 m of pipeline (19.46 m; 5.4 m/100 m).
2. A pipeline 2.5 km long and 150 mm diameter supplies water from a reservoir to a small town storage tank. Calculate the discharge when the pipe outlet is freely discharging into the tank and the difference in level between the reservoir and the outlet is 15 m. Assume $\lambda = 0.04$ (0.012 m³/s).

3. Water from a large reservoir flows through a pipeline, 1.8 km long and discharges into service tank. The first 600 m of pipe is 300 mm in diameter and the remainder is 150 mm in diameter. Calculate the discharge when the difference in water level between the two reservoirs is 25 m and $\lambda = 0.04$ for both pipes (0.02 m³/s).

4. A venturi meter is fitted to a pipeline to measure discharge. The pipe diameter is 300 mm and the venturi throat diameter is 75 mm. Calculate the discharge in m³/s when the difference in pressure between the pipe and the venturi throat is 400 mm of water. Assume the coefficient of discharge is 0.97 (0.07 m³/s).

5. A pipeline is reduced in diameter from 500 to 300 mm using a concentric reducer pipe. Calculate the force on the reducer when the discharge is 0.35 m³/s and the pressure in the 500 mm pipe is 300 kN/m² (37.4 kN).

6. A 500-mm diameter pipeline is fitted with a 90° bend. Calculate the resultant force on the bend and its line of action if the normal operating pressure is 50 m head of water and the discharge is 0.3 m³/s. Calculate the resultant force when there is no flow in the pipe but the system is still under pressure (138 kN; 137 kN).

7. Calculate the pressure rise in a 0.5 m diameter pipeline carrying a discharge of 0.3 m³/s when a sluice valve is closed suddenly (187 m).

Chapter 5

Channels

5.1 INTRODUCTION

Natural rivers and man-made canals are open channels. They have many advantages over pipes and have been used for many centuries for water supply, transport and agriculture. The Romans made extensive use of channels and built aqueducts for their sophisticated water supply schemes. Barge canals are still an important means of transporting heavy bulk materials in Europe and irrigation canals bring life and prosperity to the arid lands of North Africa, Middle East, India and Australia as they have done for thousands of years.

5.2 PIPES OR CHANNELS?

The choice between pipes and open channels is most likely to depend on which provides the cheapest solution both in terms of capital expenditure and the recurrent costs of operation and maintenance. However, there are advantages and disadvantages with each which may influence or restrict the final choice.

Channels, for example, are very convenient and economical for conveying large quantities of water over relatively flat land such as in large irrigation systems on river flood plains. It is hard to imagine some of the large irrigation canals in India and Pakistan being put into pipes, although 5 m diameter pipes are used in Libya to transport water many hundreds of kilometres across the desert to coastal cities and irrigation schemes. In hilly areas, the cost of open channels can rise significantly because the alignment must follow the land contours to create a gentle downward slope for the flow. A more direct route would be too steep causing erosion and serious damage to channels. Pipes would be more suitable in such conditions. They can be used in any kind of terrain and can take a more direct route. Water velocities too can be much higher in pipes because there is no risk of erosion.

Although there are obvious physical differences between channels and pipes, there are several important hydraulic differences as well:

- Water can only flow downhill in channels but in pipes it can flow both uphill and downhill. Flow in pipes depends on a pressure difference between the inlet and outlet. As long as the pressure is higher at the inlet than at the outlet then water will flow even though the pipeline route may be undulating. Channels depend entirely on the force of gravity to make water move and so they can only flow downhill.
- Man-made channels can have many different shapes (circular, rectangular or trapezoidal) and sizes (different depths, widths and velocities). Natural river channels are irregular in shape (Figure 5.1). Pipes in contrast are circular in section and their shape is characterised by one simple dimension – the diameter. This fixes the area of the waterway and the friction from the pipe circumference.
- Water velocity is usually lower in channels than in pipes. This is because channels are often in natural soils which erode easily. So channels are usually much larger than pipes for the same flow.
- Channels need much more attention than pipes. As they tend to erode and weeds grow in waterways, regular cleaning is required. Water losses from seepage and evaporation can also be problem.

These differences make channels a little more complicated to deal with than pipes but most open channel problems can be solved using the basic tools of hydraulics; discharge and continuity, energy and momentum.

The study of open channels is not just confined to channel shapes and sizes. It can also include waves; the problem of handling varying flood flows down rivers and sediment transport associated with the scouring and silting of rivers and canals. Some of these issues are touched on in this chapter.

5.3 LAMINAR AND TURBULENT FLOW

Laminar and turbulent flow both occur in channels as well as in pipes. But for all practical purposes, flow in rivers and canals is turbulent and, like pipes, laminar flow is unlikely to occur except for very special conditions. For example, laminar flow only occurs in channels when the depth is less than 25 mm and the velocity is less than 0.025 m/s. This is not a very practical size and velocity and so laminar flow in channels can safely be ignored. The only time it does become important is in laboratory studies where physical hydraulic models are used to simulate large and complex channel flow problems. Scaling down the size to fit in the laboratory often means that the flow in the model becomes laminar. This change in flow

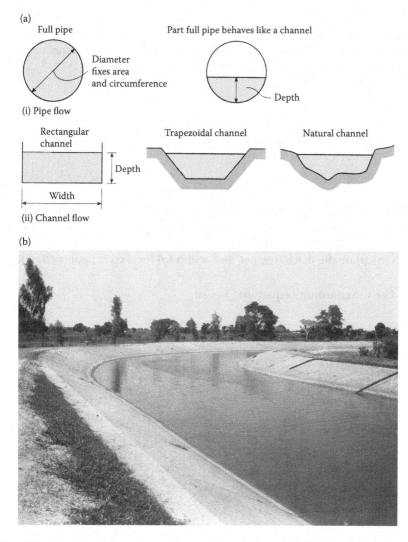

(a)

Full pipe

Part full pipe behaves like a channel

Diameter
fixes area
and circumference

Depth

(i) Pipe flow

Rectangular
channel

Trapezoidal channel

Natural channel

Depth

Width

(ii) Channel flow

(b)

Figure 5.1 Channels can have many different shapes and sizes. (a) (i) Full and part full pipe flow. (ii) Different channel shapes. (b) A lined irrigation canal in Nigeria.

regime will affect the results and care is needed when using them to assess what will happen in real situation.

5.4 USING THE HYDRAULIC TOOLS

Continuity and energy are particularly useful tools for solving open channel flow problems. Momentum is also helpful for problems in which there are energy losses and where there are forces involved.

5.4.1 Continuity

Continuity is used for open channels in much the same way as it is used for pipes (Figure 5.2a). The discharge Q_1 passing point 1 in a channel must be equal to the discharge Q_2 passing point 2.

$$Q_1 = Q_2.$$

Writing this in terms of velocity and area

$$v_1 a_1 = v_2 a_2.$$

The term discharge per unit width (q) is often used to describe channel flow rather than the total discharge (Q). This is the flow in a 1.0 m wide portion of a channel (Figure 5.2b).

To calculate the discharge per unit width (q) for a rectangular channel

Use the continuity equation $Q = va$,

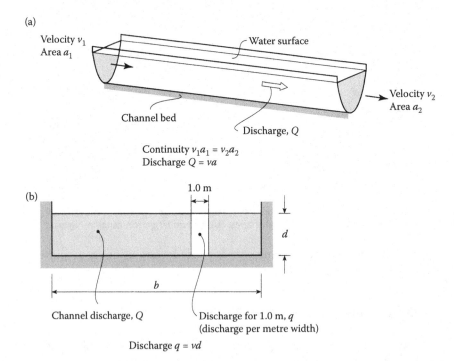

Figure 5.2 Continuity in channels. (a) Continuity along a channel. (b) Unit discharge.

where

$$a = bd.$$

To calculate q assume $b = 1.0$ m. Therefore

$$a = 1.0 \times d = d,$$

and so

$$q = vd.$$

So when a channel width (b) is 7 m and it carries a discharge (Q) of 10 m³/s, the flow per unit width is calculated as follows

$$q = \frac{Q}{b}$$

$$= \frac{10}{7} = 1.43 \text{ m}^3\text{/s/m width of channel.}$$

The discharge and continuity equations for channels are often written in terms of discharge per unit width as follows

$$q = vd,$$

$$v_1 d_1 = v_2 d_2.$$

5.4.2 Energy

The idea of total energy being the same at different points in a system is also useful for channel flow. So when water flows between two points 1 and 2 in a channel, the total energy at 1 will be the same as the total energy at 2. As in the case of pipe flow this is only true when there is no energy loss. For some channel problems this is a reasonable assumption to make but for others an additional energy loss term is needed.

The energy equation for channel flow is a little different to the equation for pipe flow (Figure 5.3). For pipe flow, the pressure energy is $p/\rho g$. For channel flow, this term is replaced by the depth of water d. Remember that $p/\rho g$ is a pressure head and is already measured in metres and so for channels this is the same as the pressure on the channel bed resulting from the depth of water d. The potential energy z is measured from some datum point to the bed of the channel. The velocity energy $v^2/2g$ remains the same.

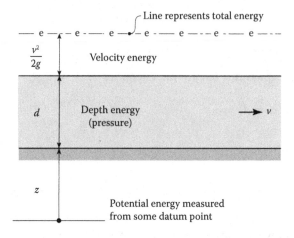

Figure 5.3 Energy in channels.

Note that, all the terms are measured in metres and so they can all be added together to determine the total energy in a channel.

Writing the total energy equation for an open channel

$$\text{Total energy} = d + \frac{v^2}{2g} + z.$$

Sometimes the velocity v is written in terms of the discharge per unit width q and depth d. This is done using the discharge equation

$$q = vd.$$

Rearrange this for velocity

$$v = \frac{q}{d}.$$

So the velocity energy becomes

$$\frac{v^2}{2g} = \frac{q^2}{2gd^2}.$$

Substitute this in the energy equation

$$\text{Total energy} = d + \frac{q^2}{2gd^2} + z.$$

Total energy can be represented diagrammatically by the *total energy line* e—e—e (Figure 5.3). This provides a visual indication of the total energy available and how it is changing. Notice that the line only slopes downwards in the direction of the flow to show the gradual loss of energy from friction. The water surface is the channel equivalent of the hydraulic gradient for pipes. It represents the pressure on the bed of the channel.

5.4.3 Using energy and continuity

One example of the use of the continuity and energy equations in an open channel is to calculate the discharge under a sluice gate (Figure 5.4a).

(a)

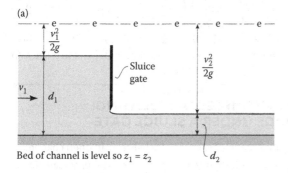

(b)

Figure 5.4 Using energy and continuity. (a) 'Free' flow under a sluice gate. (b) Sluice gates controlling discharge into irrigation canal, Iraq.

The sluice gate is a common structure for controlling flows in channels and it can also be used to measure flow if the water depths upstream and downstream of the gate are measured. The approach is very similar to the venturi problem in pipe flow but in this case a gate is used to change the energy conditions in the channel. Notice how the energy line has been drawn to indicate the level of total energy. Firstly, it shows there is no energy loss as water flows under the gate. This is reasonable because the flow is converging under the gate and this tends to suppress turbulence which means little or no energy loss. Secondly, it shows that there is a significant change in the components of the total energy across the gate even though the total is the same. Upstream the flow is slow and deep whereas downstream the flow is very shallow and fast. The discharge is the same on both sides of the gate but it is clear that the two flows are quite different. In fact they behave quite differently too, but more about this in Section 5.7.

The example in Box 5.1 illustrates how to calculate the discharge under a sluice gate when the upstream and downstream water depths are known.

BOX 5.1   EXAMPLE: CALCULATING DISCHARGE UNDER A SLUICE GATE

A sluice gate is used to control and measure the discharge in an open channel. Calculate the discharge per unit width in the channel when the upstream and downstream water depths are 1.0 and 0.2 m, respectively (Figure 5.5).

When the flow is contracting, as it does under a sluice gate, turbulence is suppressed and the flow transition occurs smoothly. Very little energy is lost and so the energy equation can be applied as follows

total energy at point 1 = total energy at point 2,

$$d_1 + \frac{v_1^2}{2g} + z_1 = d_2 + \frac{v_2^2}{2g} + z_1.$$

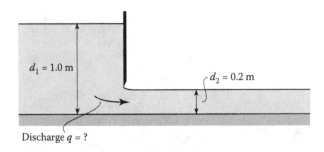

Figure 5.5 Calculating discharge under a sluice gate.

As the channel is horizontal

$$z_1 = z_2,$$

and so

$$d_1 + \frac{v_1^2}{2g} = d_2 + \frac{v_2^2}{2g}.$$

Bring the d terms and v terms together and put in the values for depth

$$1.0 - 0.2 = \frac{v_2^2}{2g} - \frac{v_1^2}{2g}.$$

Both velocities v_1 and v_2 are unknown and so the continuity equation is needed to solve the problem

$$v_1 d_1 = v_2 d_2.$$

Put in the depths

$$v_1 1.0 = v_2 0.2,$$

and so

$$v_1 = 0.2 v_2.$$

Substitute for v_1 in the energy equation

$$0.8 = \frac{v_2^2}{2g} - 0.04 \frac{v_2^2}{2g}.$$

Rearrange this to find v_2

$$v_2^2 = \frac{0.8 \times 2 \times 9.81}{1 - 0.04} = 16.35,$$

$$v_2 = 4 \text{ m/s}.$$

Calculate the discharge

$$q = v_2 d_2,$$

$$q = 4 \times 0.2,$$

$$q = 0.8 \text{ m}^3/\text{s/m width of channel}.$$

5.4.4 Taking account of energy losses

Energy loss occurs in channels due to friction. In short lengths of channel, such as in the sluice gate example, energy loss is very small and so it is not taken into account in any calculations. But energy loss in long channels must be taken into account in the energy equation to avoid serious errors (Figure 5.6). Writing the energy equation for two points in a channel

total energy at 1 (TE_1) = total energy at 2 (TE_2) + h_f,

$$d_1 + \frac{v_1^2}{2g} + z_1 = d_2 + \frac{v_2^2}{2g} + z_2 + h_f,$$

where h_f is energy loss due to friction (m).

This equation is very similar to that for pipe flow and in that case the Darcy–Weisbach formula was used to calculate the energy loss term h_f. This was the link between energy losses and pipe size. Similar formulae have been developed to calculate h_f for open channels and these link energy losses both to the size and shape of channels needed to carry a given discharge. But as with pipe flow there are a few important steps to be taken before getting to the formula. The first of these steps is the concept of uniform flow.

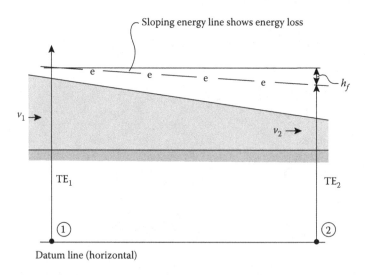

Figure 5.6 Energy loss in channels.

5.5 UNIFORM FLOW

All the energy loss formulae for open channels are based on *uniform flow*. This is a special condition that only occurs when water flows down a long, straight, gently sloping channel (Figure 5.7). The flow is pulled down the slope by the force of gravity but there is friction from the bed and sides of the channel slowing it down. When the friction force is larger than the gravity force it slows down the flow. When the friction force is smaller than the gravity force the flow moves faster down the slope. But the friction force is not constant, it depends on velocity and so as the velocity increases so does the friction force. At some point the two forces become equal. Here the forces are in balance and as the flow continues down the channel, the depth and velocity remain constant. This flow condition is called *uniform flow* and the water depth is called the *normal depth*.

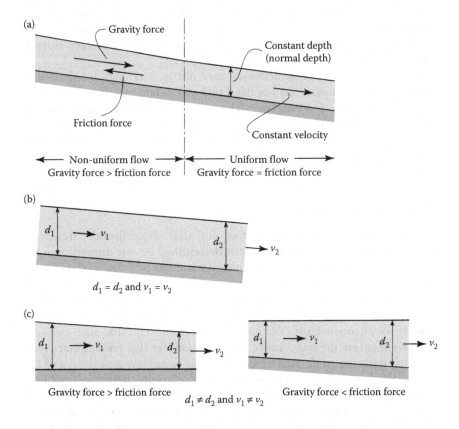

Figure 5.7 Uniform and non-uniform flow. (a) Uniform flow occurs in a long, gently sloping channel. (b) Uniform flow – gravity force = friction force. (c) Non-uniform flow.

Unfortunately, in most channels this balance of forces rarely occurs and so the depth and velocity are usually changing gradually even though the discharge is constant. Even in long channels where uniform flow has a chance of occurring there is usually some variation in channel shape or slope or a hydraulic structure which changes the depth and the velocity (Figure 5.7c). So most channels have *non-uniform flow*. It is also called *gradually varied flow* because the changes take place gradually along the channel.

So if non-uniform flow is the most common and uniform flow rarely occurs, why are all the formulae based on uniform flow? Why not accept that non-uniform flow is the norm and develop formulae for this condition? The answer is quite simple. Uniform flow is much easier to deal with from a calculation point of view and engineers are always looking for ways of simplifying problems but without losing accuracy. There are methods for designing channels for non-uniform flow but they are much more cumbersome to use and usually they produce the same shape and size of channel as uniform flow methods. So it has become accepted practice to assume that channel flow is uniform for all practical purposes. For all gently sloping channels on flood plains (i.e. 99% of all channels including rivers, canals and drainage ditches) this assumption is a good one. Only in steep sloping channels in the mountains does it cause problems.

Remember too that great accuracy may not be necessary for designing channels and dimensions do not need to be calculated to the nearest millimetre for construction purposes. The nearest 0.05 m is accurate enough for concrete channels and probably 0.1 m for earth channels is more than enough. From a practical construction point of view it will be difficult to find a hydraulic excavator operator who can (or who would want to) trim channel shapes to an accuracy greater than this.

5.5.1 Channel shapes

For pipe flow the issue of shape does not arise. Pipes are all circular in section and so their hydraulic shape is determined by one dimension – the pipe diameter – which takes account of both shape and size in the Darcy–Weisbach formula. But channels come in a variety of shapes, the more common ones being rectangular, trapezoidal or semicircular (Figure 5.8). They also come in different sizes with different depths and widths. So any formula for channels must take into account both shape and size.

To demonstrate the importance of shape consider two rectangular channels each with the same area of flow but one is narrow and deep and the other is shallow and wide (Figure 5.8a). Both channels have the same flow area of 10 m² and so they might be expected to carry the same discharge. But this is not the case. Friction controls the velocity in channels and when the friction changes the velocity will also change. The channel boundary in contact with the water (called the *wetted perimeter*) is the main source of friction. In the narrow channel the length of the boundary in contact with

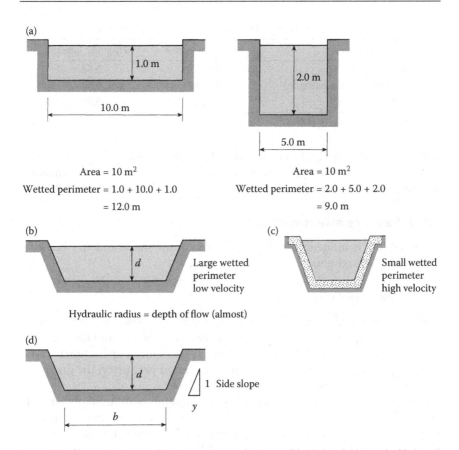

Figure 5.8 Channel shapes. (a) Importance of shape. (b) Unlined channel. (c) Lined channel. (d) Trapezoidal channel section.

the water is 9 m, whereas in the wide channel it is 12 m. So the wide channel produces more friction than the narrow one and as a result the velocity (and hence the discharge) in the wide channel will be less than in the narrow one. So channels with the same flow area have different carrying capacities depending on their shape.

Understanding this can be very useful when deciding on the general shape of channels. Suppose you are constructing a channel in natural soil and there are worries about erosion of the bed and sides because of a high water velocity. By making the channel wide and shallow the increased friction will slow down the water and avoid the problem (Figure 5.8b). Many natural river channels have a shallow, wide profile as they have adapted over many years to the erosion of the natural soils in which they flow.

For lined channels (e.g. concrete), the main issue is cost and not erosion. Lined canals are expensive and so it is important to minimise the amount of lining needed. This is done by using a channel shape that has a small wetted

Table 5.1 Minimum wetted perimeters for different channel shapes

Channel shape	Wetted perimeter	Hydraulic radius
Rectangle	4d	0.5d
Trapezoid (half a hexagon)	3.463d	0.5d
Semicircle	0.5πd	0.5d

Note: *d* is the depth of flow.

perimeter (Figure 5.8c). Friction will be low which means the velocity will be high but this is not a problem as the lining will resist erosion. Minimum wetted perimeters for selected channel shapes are shown in Table 5.1.

5.5.2 Factors affecting flow

Channel flow is influenced not just by the channel shape and size but also by slope and roughness.

5.5.2.1 Area and wetted perimeter

The cross sectional area of a channel (*a*) defines the flow area and the wetted perimeter (*p*) defines the boundary between the water and the channel. This boundary in contact with the water is the source of frictional resistance to the flow of water. The greater the wetted perimeter, the greater is the frictional resistance of the channel.

The area and wetted perimeter for rectangular and circular channels are easily calculated but trapezoidal channels are a bit more difficult. Unfortunately, they are the most common and so given below are formulae for area and wetted perimeter (Figure 5.8d).

Area of waterway $(a) = (b + yd)d,$

Wetted perimeter $(p) = b + 2d(1 + y^2)^{1/2},$

where *b* is bed width (m), *d* is depth of flow (m) and *y* is side slope.

5.5.2.2 Hydraulic radius

As the wetted perimeter can vary considerably for the same area, some measure of the hydraulic shape of a channel is needed. This is called the *hydraulic radius* (R) and it is determined from the area and the wetted perimeter as follows

$$\text{Hydraulic radius (m)} = \frac{\text{area (m}^2)}{\text{wetted perimeter (m)}}.$$

In the two channels in Figure 5.8, the hydraulic radius would be 1.11 and 0.83 m, respectively, and this shows numerically just how hydraulically different are the two channels.

5.5.2.3 Slope

Water only flows downhill in channels and the steepness of the slope affects the velocity, and hence the discharge. As the slope gets steeper the velocity increases and so does the discharge. (Remember $Q = va$).

Slope is measured as a gradient rather than an angle in degrees. So a channel slope is expressed as 1 in 1,000, that is, 1.0 m drop in 1,000 m of channel length (Figure 5.9).

Slopes need not be very steep for water to flow. Many of the irrigation canals in the Nile valley in Egypt have slopes of only 1 in 10,000. This is the same as 1.0 m in 10 km or 100 mm/km. This is a very gentle slope and you would not be able to detect it by just looking at the landscape, but it is sufficient to make water flow as the evidence of the Nile shows.

A question which sometimes arises about channel slopes is: Does slope refer to the water surface slope, the channel bed or the slope of the energy line? For uniform flow the question is irrelevant because the depth and velocity remain the same along the entire channel and so the water surface and the bed are parallel and have the same slope. For non-uniform flow it is the slope of the energy line that is important and is the driving force for the flow. Even when the bed is flat, water will still flow provided there is an energy gradient.

5.5.2.4 Roughness

The *roughness* of the bed and sides of a channel also contribute to friction. The rougher they are, the slower will be the water velocity. Channels tend to have much rougher surfaces than pipes. They may be relatively smooth when lined with concrete but they can be very rough when excavated in the natural soil or infested with weeds. Roughness is taken into account in channel design formula and this is demonstrated in the next section.

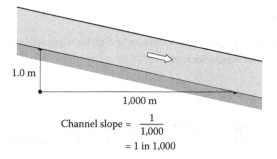

Figure 5.9 Channel slope.

5.5.3 Channel design formulae

There are two commonly used formulae which link energy loss in channels to their size, shape, slope and roughness; the *Chezy* formula and the *Manning* formula. Both are widely used and were developed on the assumption that the flow is uniform.

5.5.3.1 Chezy formula

This formula was developed by Antoine Chezy, a French engineer who, in 1768, was asked to design a canal for the Paris water supply. The Chezy formula, as it is now known, is usually written as follows

$$v = C\sqrt{RS},$$

where v is velocity (m/s), R is hydraulic radius (m), S is channel slope (m/m) and C is the Chezy coefficient describing channel roughness.

This may not look much like a formula for friction loss in a channel but it is derived from the energy equation allowing for energy loss in the same way as was done for pipe flow. For pipe flow the outcome was the Darcy–Weisbach formula, for channel flow the outcome is the Chezy formula. For those interested in the origins of this formula, a derivation is shown in Box 5.2. The formula shown above is the more familiar way of presenting the Chezy formula in hydraulic textbooks.

Once the velocity has been calculated, the discharge can be determined using the discharge equation $Q = va$.

BOX 5.2   DERIVATION: CHEZY FORMULA

To see how Chezy developed his formula, look first at the energy equation describing the changes which take place between two points in a long channel (Figure 5.7b) and taking into account energy loss in the channel h_f.

$$d_1 + \frac{v_1^2}{2g} + z_1 = d_2 + \frac{v_2^2}{2g} + z_2 + h_f.$$

Chezy also studied pipe flow formulae and so he suggested that the energy loss h_f in channels would format

$$h_f = \frac{Lv^2}{C^2R},$$

where L is the length of channel over which the energy loss occurs (m), v is velocity (m/s), R is hydraulic radius (m) and C is Chezy coefficient describing roughness.

Notice how similar this equation is to the Darcy–Weisbach equation. Friction depends on the length and the square of the velocity. The Chezy coefficient C describes the friction in the channel and is similar to λ in the Darcy–Weisbach formula. Like λ it does not have a constant value. C depends on the Reynolds number and also on the dimensions of the channel.

Put this into the energy equation

$$d_1 + \frac{v_1^2}{2g} + z_1 = d_2 + \frac{v_2^2}{2g} + z_2 + \frac{Lv^2}{C^2R}.$$

Now uniform flow is defined by the depths and velocities remaining the same along the whole length of a channel and so

$$d_1 = d_2,$$

and

$$v_1 = v_2.$$

This reduces the energy equation to

$$\frac{Lv^2}{C^2R} = z_1 - z_2.$$

Now divide both sides of the equation by L

$$\frac{v^2}{C^2R} = \frac{z_1 - z_2}{L}.$$

But $\frac{z_1 - z_2}{L}$ is the slope of the channel bed S. It is also the slope of the water surface. Remember that the two are parallel for uniform flow. Hence

$$\frac{z_1 - z_2}{L} = S,$$

and so

$$\frac{v^2}{C^2R} = S.$$

Rearrange this to calculate the velocity

$$v = C\sqrt{RS}.$$

This is the familiar form of the Chezy equation that is quoted in hydraulic text books.

5.5.3.2 Manning formula

The Manning formula is an alternative to Chezy and is one of the most commonly used formulae for designing channels. This was developed by Robert Manning (1816–1897), an Irish civil engineer. It is an empirical formula developed from many observations made on natural channels.

$$v = \frac{R^{2/3}S^{1/2}}{n},$$

where v is velocity (m/s), R is hydraulic radius (m), S is channel bed slope (m/m) and n is Manning's roughness coefficient. Manning's n values depend on the surface roughness of the channel bed and sides. Typical values are listed in Table 5.2.

The value of Manning's n is not just determined by the material from which the channel is made but it is also affected by vegetation growth. This can make it difficult to determine with any accuracy. The n value can also change over time as weeds grow and it can also change with changes in flow. At low discharges weeds and grasses will be upright and so cause great roughness but at higher discharges they may be flattened by the flow and so the channel becomes much smoother. There is an excellent book, *Open Channel Flow* by Ven Te Chow (see Further Reading) which has a series of pictures of channels with different weed growths and suggested n values. These pictures can be compared with existing channels to get some indication of n. But how is Manning's n selected for a natural, winding channel with varying flow areas; with trees and grasses along its banks (perhaps also including the odd bicycle or supermarket trolley) and flowing under bridges and over weirs? Clearly, in this situation, choosing n is more of an art than a science. It may well be that several values are needed to describe the roughness along different sections of the river.

5.5.4 Using Manning's formula

Manning's formula is not the easiest of formula to work with. It is quite straightforward to use when calculating discharge for a given shape and

Table 5.2 Values of Manning's n

Channel type	Manning's n values
Concrete lined canals	0.012–0.017
Rough masonry	0.017–0.030
Roughly dug earth canals	0.025–0.033
Smooth earth canals	0.017–0.025
Natural river in gravel	0.040–0.070

size of channel but it is not so easy to use to the other way round, that is, to calculate channel dimensions for a given discharge. Unfortunately, this is by far the most common use of Manning. One approach is to use a trial and error technique to obtain the channel dimensions. This means guessing suitable values for depth and width and then putting them into the formula to see if they meet the discharge requirements. If they do not then the values are changed until the right dimensions are found. Usually there can be a lot of trials and a lot of errors. Modern computer spreadsheets can speed up this painful process.

Another approach for those who do not have spreadsheet skills is the method developed by HW King in his *Handbook for the Solution of Hydrostatic and Fluid Flow Problems* (see Further Reading). This is a simple and useful method and is ideally suited to designing trapezoidal channels, which are the most common. King modified Manning's formula to look like this:

$$Q = \frac{1}{n} jk d^{8/3} S^{1/2},$$

where d is depth of flow (m), S is channel slope and j and k are constants. The values of j and k depend on the ratio of the bed width to depth and the channel side slope. This is the slope of the side embankments and not the longitudinal slope of a channel S. King's book has a very comprehensive range of j and k values. A selection of the most common values is shown in Table 5.3.

To use the method, values of the ratio of channel bed width to depth and side slope are first chosen. Values of j and k are then obtained from Table 5.3 and put into the formula from which a value of depth can be calculated. As the bed width to depth ratio is known, the bed width can now be calculated. If the resulting channel shape or its dimensions appear to be unsuitable for any reason (e.g. the velocity may be too high) then another ratio of bed width to depth ratio can be chosen and the calculation repeated. As well as providing j and k values, King also supplies values

Table 5.3 Values of *j* and *k* for Manning's formula

	Ratios of bed width (b) to water depth (d)					
	b = d	b = 2d	b = 3d	b = d	b = 2d	b = 3d
Side slope	Values of j			Values of k		
Vertical	1	2	3	0.48	0.63	0.71
1 in 1	2	3	4	0.64	0.73	0.77
1 in 1.5	2.5	3.5	4.5	0.66	0.73	0.77
1 in 2	3	4	5	0.66	0.72	0.76
1 in 3	4	5	6	0.67	0.71	0.74

of the power function 8/3 so that it is easy to calculate the depth of flow. Remember that his book was written some years ago before everyone had a calculator.

Examples using Manning's formula are in Boxes 5.3 and 5.4.

BOX 5.3   EXAMPLE: CALCULATING DISCHARGE USING MANNING'S FORMULA

Calculate the discharge in a rectangular concrete lined channel of width 2.5 m and depth 0.5 m with a slope of 1 in 2,000 and a Manning's n value is 0.015 (Figure 5.10).

The first step is to calculate the velocity but before this can be done, the area, wetted perimeter and hydraulic radius must be determined:

$$\text{Area } (a) = \text{depth} \times \text{width}$$

$$= 0.5 \times 2.5 = 1.25 \text{ m}^2.$$

$$\text{Wetted perimeter } (p) = 0.5 + 2.5 + 0.5 = 3.5 \text{ m},$$

and so

$$\text{Hydraulic radius } (R) = \frac{a}{p} = \frac{1.25}{3.5} = 0.36 \text{ m}.$$

Next calculate velocity

$$v = \frac{0.36^{2/3} \times \left(\frac{1}{2,000} \right)^{1/2}}{0.015} = 0.75 \text{ m/s}.$$

Now calculate discharge

$$Q = va,$$

$$Q = 1.25 \times 0.75 = 0.94 \text{ m}^3/\text{s}.$$

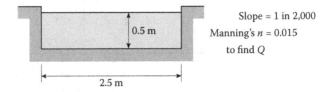

Slope = 1 in 2,000
Manning's n = 0.015
to find Q

0.5 m

2.5 m

Figure 5.10 Calculating discharge using Manning's equation.

BOX 5.4   EXAMPLE: CALCULATING DEPTH OF FLOW AND BED WIDTH USING MANNING FORMULA (KING'S METHOD)

Calculate a suitable bed width and depth of flow for an unlined trapezoidal channel to carry a discharge of 0.6 m³/s on a land slope of 1 in 1,000 (Figure 5.11). The soil is a clay loam and so the side slope will be stable at 1.5:1 and the maximum permissible velocity is 0.8 m/s (Table 5.4).

The first step is to select a suitable value for Manning's n ($n = 0.025$ for natural soil) and then select a bed width to depth ratio (try $b = d$).

Now obtain values for j and k from Table 5.3, that is, $j = 2.5$ and $k = 0.66$. Calculate d using the Manning formula

$$Q = \frac{1}{n} jkd^{8/3}s^{1/2},$$

$$0.6 = \frac{1}{0.025} \times 2.5 \times 0.66 \times d^{8/3} \times 0.001^{1/2}.$$

Rearrange this for d

$$d^{8/3} = 0.286,$$

$$d = 0.63 \text{ m}.$$

As the ratio of bed width to depth is known, calculate b. In this case

$$b = d,$$

and so

$$b = 0.63 \text{ m}.$$

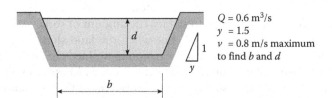

$Q = 0.6$ m³/s
$y = 1.5$
$v = 0.8$ m/s maximum
to find b and d

Figure 5.11 Calculating depth of flow and bed width using King's method.

All the channel dimensions are now known but do they comply with the velocity limit (Table 5.4)? Check the velocity using the discharge equation

$Q = av$.

For a trapezoidal channel

$a = (b + yd)d$

$= (0.63 + 1.5 \times 0.63)\, 0.63 = 0.993 \text{ m}^2$.

Substitute this and the value for discharge into the discharge equation

$0.6 = 0.993 \times v$.

Calculate velocity

$v = 0.6 \text{ m/s}$.

This is less than the maximum permissible velocity of 0.8 m/s and so these channel dimensions are acceptable.

Note that there are many different channel dimensions that could be chosen to meet the design criteria. This is just one answer. Choosing another $b{:}d$ ratio would produce different dimensions but they would be acceptable provided they met the criteria. Increasing the $b{:}d$ ratio would reduce the velocity whereas decreasing the $b{:}d$ ratio would increase the velocity. The latter would not be an option in this example as the velocity is close to the maximum permissible already.

A freeboard would normally be added to this to ensure that the channel is not over-topped.

5.5.5 Practical design

In engineering practice, the usual design problem is to determine the size, shape and slope of a channel to carry a given discharge. There are many ways to approach this problem but here are some guidelines.

Whenever possible, the channel slope should follow the natural land slope. This is done for cost as it helps to reduce the amount of soil excavation and embankment construction needed. But when the land slope is steep, high water velocities may occur and cause erosion in unlined channels. The most effective way to avoid erosion is to limit the velocity. Maximum non-scouring velocities for different soil types are shown in Table 5.4. Channels can also be lined for protection. Slope can also be reduced to lower the velocity to an acceptable level by using drop structures to take the flow down the slope in a series of steps – like a staircase (see Section 6.10).

Table 5.4 Maximum permissible velocities

Material	Maximum velocity (m/s)
Silty sand	0.3
sandy loam	0.5
Silt loam	0.6
Clay loam	0.8
Stiff clay	1.1

It is important to understand that there is no single correct answer to the size and shape of a channel, but a range of possibilities. If three people were each asked to design a trapezoidal channel for a given discharge, it is likely that they would come up with three different answers, and all could be acceptable. It is the designer's job to select the most appropriate one. Usually, the selection is made simpler because of the limited range of values that are practicable. For example, land slope will limit the choice of slope and the construction materials will limit the velocity. But even within these boundaries there are still many possibilities.

One of the problems of channel design is that of choosing suitable values of depth and bed width. King's method gets around this problem by asking the designer to select a ratio between them rather than the values themselves. Another way to simplify the problem is to assume that the hydraulic radius R is equal to the depth of the water d. This is a reasonable assumption to make when the channel is shallow and wide. Referring to the example in Figure 5.8a, the hydraulic radius was 1.11 m in the wider channel when the depth was 1.0 m. This is close enough for channel design purposes. The depth can then be calculated using the Manning formula and the bed width determined using the area and discharge.

The depth and width of a channel also influence velocity. For lined channels, which are expensive, it is important to keep the wetted perimeter (p) as small as possible as this keeps the cost down. This results in channels which are narrow and deep (Figure 5.8c). For unlined channels, the velocity must be kept well within the limits set in Table 5.4. Making channels wide and shallow increases the wetted perimeter and channel resistance and this slows down the flow (Figure 5.8b). Look at any stream or river flowing in natural soil. Unless it is constrained by rocks or special training works, it will naturally flow wide and shallow. So new channels which are to be constructed in similar material should also follow this trend.

5.6 NON-UNIFORM FLOW: GRADUALLY VARIED

There are two kinds of non-uniform flow. The first is *gradually varied flow*. This is the most common type of flow and has already been described

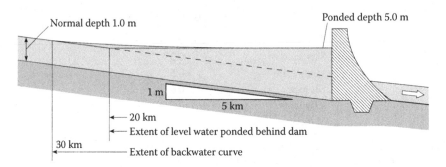

Figure 5.12 Backwater curve.

earlier in this chapter. It occurs when there are gradual changes taking place in the depth and velocity due to an imbalance of the force of gravity trying to make the flow go faster down a slope and the channel friction slowing it down. The gradual changes in depth take place over long distances and the water surface follows a gradual curve.

Engineers recognise 12 different surface water curves depending on the different gradually varied flow conditions that can occur in channels but the most common is the *back-water curve*. This occurs when a channel is dammed (Figure 5.12). For example, a river flowing at a normal depth of 1.0 m down a gradient of 1 m in 5 km is dammed so that the water level rises to a depth of 5.0 m. For a level water surface behind the dam, its influence extends 20 km upstream. But because the river is flowing there is a backwater curve which extends the influence of the dam up to 30 km. This effect can be important for river engineers who wish to ensure that a river's embankments are high enough to contain flows and for landowners along a river whose land may be flooded by the dam construction. The backwater curve can be predicted using the basic tools of hydraulics but they go beyond the scope of this book. One problem is that they depend largely for their accuracy on predicting the value of Manning's *n* which can be very difficult in natural channels.

5.7 NON-UNIFORM FLOW: RAPIDLY VARIED

The second type of non-uniform flow is *rapidly varied flow*. As its name implies, sudden changes in depth and velocity occur and this is the result of sudden changes in either the shape or size of channels. The change usually takes place over a few metres, unlike gradually varied flow where changes take place slowly over many kilometres. Hydraulic structures are often the cause of rapidly varied flow and the sluice gate in Section 5.4 is a good example of this. In this case the gate changed the flow suddenly from a deep, slow flow upstream to a fast, shallow flow downstream. Building a weir or widening (or deepening) a channel will also cause sudden changes to occur. But unfortunately, all flows do not behave in the same way. For example, a weir in a channel will have quite a different effect on the deep, slow flow than on the shallow, fast flow. So a further classification of channel flow is needed, this time in terms of how flow behaves when channel size or shape is changed suddenly.

The two contrasting types of flow described above are now well recognised by engineers. The more scientific name for deep, slow flow is *subcritical flow* and for shallow, fast flow is *supercritical flow*. This implies that there is some *critical point* when the flow changes from one to the other and that this point defines the difference between the two flow types. This is indeed the case. At the critical point the velocity becomes the *critical velocity* and the depth becomes the *critical depth*. The critical point is important not just to classify the two flow types but it also plays an important role in measuring the discharges in channels. This is discussed later in Chapter 6.

5.7.1 Flow behaviour

Just how do subcritical and supercritical flows behave when there are sudden changes in the channel?

5.7.1.1 Subcritical flow

This is by far the most common flow type and is associated with all natural and gently flowing rivers and canals. The effect of a sluice gate on this kind of flow has already been described. The effect of a weir is very similar (Figure 5.13a). A weir is like a step up on the bed of a channel. Such a step causes the water level to drop and the velocity to increase. The step up reduces the flow area in a channel but the water does not slow down because of this. Its velocity increases (remember the way flow behaves in constrictions, Section 3.7.1). This increases the kinetic energy, but as there is no change in the total energy this is at the expense of the depth (pressure) energy. So the depth is reduced causing a drop in the water level. Weirs are a common sight on rivers and most people will have seen this sudden but smooth drop in water level over a weir. A similar, though not so dramatic, drop in water level occurs when water flows under a bridge. The water level drops because the reduced width of the river increases its velocity.

Interestingly, the converse is true. When a channel is made larger by increasing its width or depth, the water level rises. This is counter intuitive and not so easy to believe. But it follows from the energy equation and it actually happens in practice (Figure 5.13b). This was highlighted by a problem facing engineers who were troubled by flooding from a river flowing through a town and under the town bridge (Figure 5.13c). During stormy weather, the river level rises and reaches the underside of the bridge.

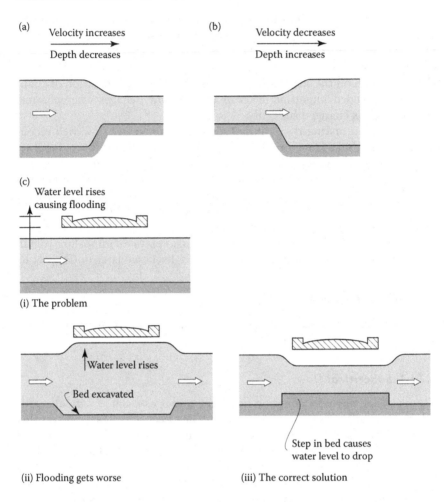

(a) Velocity increases → Depth decreases

(b) Velocity decreases → Depth increases

(c) Water level rises causing flooding

(i) The problem

Water level rises
Bed excavated

(ii) Flooding gets worse

Step in bed causes water level to drop

(iii) The correct solution

Figure 5.13 Rapidly varied flow – subcritical. (a) A step-up in a channel. (b) A step-down in a channel bed. (c) Dealing with a flooding problem.

The extra friction from the bridge slows the flow causing the water level upstream of the bridge to rise even further and flood the town. The problem was how to increase the carrying capacity of the river through the town, and particularly under the bridge, to avoid the flooding. The engineers decided that the most obvious solution was to make the channel deeper – but this made flooding worse, not better. Clearly, the engineers did not understand the hydraulic tools of continuity and energy. The increase in channel depth reduced the velocity and hence the kinetic energy. As the total energy remained the same, the depth energy increased causing the river level to rise and not fall as expected. They eventually opted for the correct solution which was to reduce the flow area under the bridge by constructing a step on the bed of the river. This increased the velocity energy and reduced the depth of flow. So even when the river was in flood, the flow was able to pass safely under the bridge. The local engineers did not believe the solution at first and insisted on building a hydraulic model in the laboratory to test it. Seeing is believing and this convinced them!

Canoeists are well aware of the way in which rapid changes in water surface levels are a direct result of changes on the river bed. They are wary of those parts of a river where the current looks swifter. It may be tempting to steer your canoe into the faster moving water but it is a sign of shallow water and there may be rocks just below the surface which can damage a canoe. The slower moving water may not be so attractive but at least it will be deep and safe.

5.7.1.2 Supercritical flow

This type of flow behaves in completely opposite way to subcritical flow. A step up on the bed of a channel in supercritical flow causes the water depth to rise as it passes over it and a channel which is excavated deeper causes the water depth to drop (Figure 5.14). Supercritical flow is very difficult to deal with in practice. Not only does the faster moving water cause severe erosion in unprotected channels but it is also difficult to control with hydraulic structures. Trying to guide a supercritical flow around a bend in a channel, for example, is like trying to drive a car at high speed around a sharp road bend. It has a tendency to over-shoot and to leave the channel. Fortunately, supercritical flows rarely occur and are confined to steep rocky streams and just downstream of sluice gates and dam spillways where water can reach speeds of 20 m/s and more. When they do occur, engineers have developed ways of quickly turning them back into subcritical flows so that they can be dealt with more easily (see Sections 5.7.6 and 6.10).

5.7.1.3 General rules

How flow behaves when there are abrupt changes to the size or shape of channels is another way of classifying channel flow:

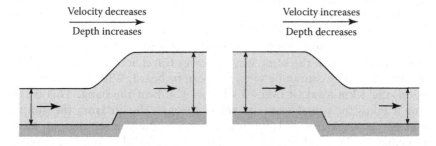

Figure 5.14 Rapidly varied flow – supercritical.

For subcritical flow, the water depth decreases when a channel flow area is reduced either by raising the bed or reducing the width. Conversely, the depth increases when the flow area is increased.

For supercritical flow, the water depth increases when the flow area is reduced and decreases when it is increased.

5.7.1.4 Spotting the difference

Sometimes the difference between subcritical and supercritical flows is obvious. The sluice gate example demonstrates both subcritical and supercritical flows in the same channel, at the same discharge and the same total energy. In this situation, there is a clear visual difference between them. But if the two flows occurred in separate straight channels, it would be more difficult to spot the difference just by looking. However, if some obstruction is put into the two flows such as a bridge pier or a sharp bend the difference between them would be immediately obvious.

A more scientific way of distinguishing between the two flows is to establish the point of change from subcritical to supercritical flow – this is the *critical point*. There are several very practical ways of determining this point. But before describing these, it would be helpful to look first at another critical point which is similar to, but perhaps more familiar than the one which occurs in channel flow.

5.7.1.5 Airflow analogy

Aeroplanes are now an everyday part of our lives. Although you may not be aware of it, and you certainly cannot see it, the airflow around an aircraft in flight is, in fact, very similar to water flow around some object in a channel. They are both fluids and so a look at airflow may help us to understand some of the complexities of water flow, and in particular the critical point.

Most people will have noticed that jumbo jets and Concorde have very different shapes (Figure 5.15). This is because the two aircraft are designed to travel at very different speeds. Jumbo jets are relatively slow and travel at only 800 km/h, whereas Concorde, a British-French supersonic passenger aircraft in service until 2003, traveled at much higher speeds of 2,000 km/h and more. But the change in aircraft shape is not a gradual one; a sudden change is needed when aircraft fly over 1,200 km/h. This is the speed at which sound waves travel through still air. Sound waves move through air in much the same way as waves travel across a water surface and although they cannot be seen, they can be heard. When someone fires a gun, say 1 km away, it takes 3 s before you hear the bang. This is the time it takes for sound waves to travel through the air from the gun to your ear at a velocity of 1,200 km/h. Notice how you see the gun flash immediately. This is because light waves travel much faster than sound waves at a velocity of 300,000 km/s. This is the reason why lightning in

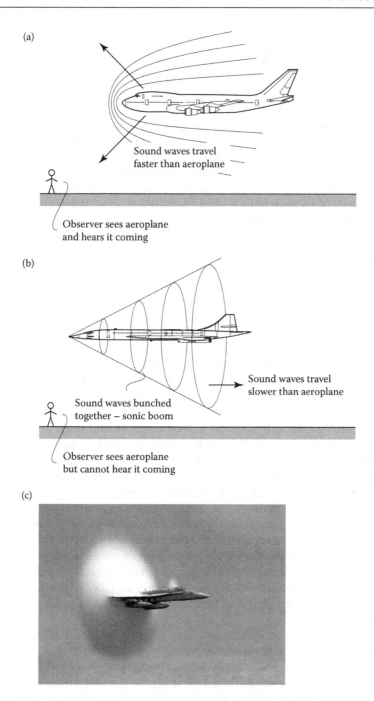

Figure 5.15 Airflow analogy. (a) Subsonic flight. (b) Supersonic flight. (c) Aircraft going through the sound barrier from subsonic to supersonic flight.

a storm is seen long before the thunder is heard even when the storm is several kilometres away.

When an aircraft is flying, the noise from its engines travels outwards in all directions in the form of sound waves. When it is travelling below the speed of sound, the sound travels faster than the aeroplane and so an observer hears the aeroplane coming before it reaches him (Figure 5.15a). This is known as *subsonic flight*. Aircraft which fly below the speed of sound have large rounded shapes like the jumbo jets. When an aeroplane is flying faster than the speed of sound, the sound travels slower than the aircraft and is left far behind. An observer will see the aircraft approaching before hearing it (Figure 5.15b). This is known as *supersonic flight*. Aircraft travelling at such speeds have slim, dart-like shapes. When the observer does eventually hear it, there is usually a loud bang. This is the result of a pressure wave, known as the *sonic boom*, which comes from all the sound waves being bunched up together behind the aeroplane.

So there are two types of airflows, subsonic and supersonic, and there is also a clear point at which the flow changes from one to the other – the speed of sound in still air.

5.7.1.6 Back to water

The purpose of this lengthy explanation about aeroplanes in flight is not just to demonstrate the close similarity between airflow and water flow but also to better explain the water phenomenon. Subcritical and supercritical flows are very similar to subsonic and supersonic flights, respectively. The majority of aircraft travel at subsonic speeds and there are very few design problems. In contrast, very few aircraft travel at supersonic speeds and their design problems are much greater and more expensive to solve. The same is true for water. Most flows are subcritical and are easily dealt with. But in a few special cases, the flow is supercritical and this is much more difficult to deal with.

The change from sub to supersonic occurs when the aeroplane reaches the speed of sound waves in still air. But this is where water is a little different. The change does not occur at the speed of sound in water, although sound does travel through water very effectively. It occurs when the flow reaches the same velocity as waves on the water surface. To avoid confusion between wave velocity and water velocity, wave velocity is often referred to as *wave celerity*.

Waves occur on the surface of water when it is disturbed. When you throw a stone into a still pond of water, waves travel out across the surface towards the bank (Figure 5.16a). Although there appears to be a definite movement of water towards the bank, it is only the waves that are moving outwards and not the water. The water only moves up and down as the waves pass. A duck floating on the pond would only bob up and down with the wave motion and would not be washed up on the bank! But the wave

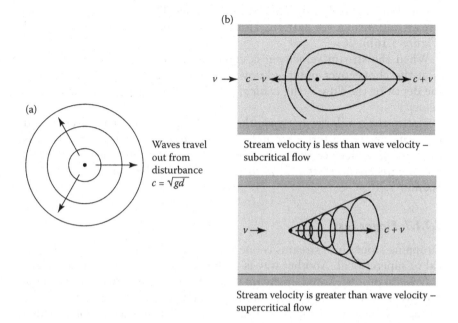

Figure 5.16 Using waves to determine flow type. (a) Waves on a still pond. (b) Waves on a flowing stream.

celerity is not a fixed value like the speed of sound. It depends on the depth of water and it can be calculated using the equation

$$c = \sqrt{gd},$$

where c is wave celerity (m/s), g is gravity constant (9.81 m/s^2) and d is depth of water (m).

Wave celerity sets the boundary between subcritical and supercritical flows in flowing water. When the flow velocity is less than the wave velocity, the flow is subcritical. When it is greater, the flow is supercritical. Let's see how this works.

As water waves are easily seen, they provide a good visual way of determining the type of flow. When there is a disturbance in a stream, waves move out in all directions but the pattern is not circular, it is distorted by the velocity of the water v (Figure 5.16b). Some waves move upstream but struggle against the flow and so appear to move more slowly than on the still pond. Others move downstream and are assisted by the flow and so they move faster. Because waves can still move upstream of the disturbance, it means that the stream velocity v is less than the wave celerity c and so the flow must be subcritical. When the stream velocity v is increased and becomes greater than the wave celerity c, then waves can no longer travel upstream against the flow. They are all swept downstream and form a vee pattern. This means the flow must be supercritical (Figure 5.16b).

When the stream velocity v is equal to the wave celerity c, the flow is at the changeover point – known as the *critical point*. At this point the depth of the flow is the *critical depth* and the velocity is the *critical velocity*.

Notice the similarity between the sound waves around an aircraft and the water wave patterns around the disturbance in a stream. There is even an equivalent of the sonic boom in water although it is much less noisy. This is the *hydraulic jump* which is described in more detail in Section 5.7.6.

5.7.1.7 Finger test

Dropping stones into streams is one way of deciding if the flow is subcritical or supercritical. Another way is to dip your finger into the water. If the waves you produce travel upstream then the flow is subcritical. If the waves are swept downstream and the water runs up your arm (you have created a stagnation point in the flow and your wet sleeve is a sign of the high velocity energy) then you are seeing supercritical flow.

5.7.2 Froude number

Another way of determining whether a flow is subcritical or supercritical is to use the Froude number. Returning to the airflow analogy for a moment, aircraft designers use the *Mach number* or *Mach speed* to describe subsonic and supersonic flights. This is a dimensionless number and is the ratio of two velocities

$$\text{Mach No.} = \frac{\text{velocity of aircraft}}{\text{velocity of sound in still air}}.$$

This dimensionless number was developed by an Austrian physicist Ernst Mach (1838–1916). A Mach No. less than 1 indicates subsonic flight and a Mach No. greater than 1 indicates supersonic flight. It follows that a Mach No. of 1 means that the aircraft is travelling at the speed of sound.

A dimensionless number similar to the Mach No. was developed by William Froude (1810–1879) to describe subcritical and supercritical flows in channels and is now referred to as the *Froude number* (F). It is a ratio of the stream velocity to the wave celerity and is calculated as follows

$$\text{Froude No.} \ (F) = \frac{\text{stream velocity (m/s)}}{\text{wave celerity (m/s)}}.$$

Note that Froude No. is dimensionless. Now wave celerity

$$c = \sqrt{gd},$$

and so

$$F = \frac{v}{\sqrt{gd}}.$$

A Froude No. less than 1 indicates subcritical flow and a Froude No. greater than 1 indicates supercritical flow. It follows that a Froude No. of 1 means that the channel is flowing at critical depth and velocity. So calculating the Froude No. is another way of determining when the flow is subcritical or supercritical.

5.7.3 Specific energy: a 'key to open a lock'

It should be possible to quantify the changes in depth and velocity resulting from abrupt changes in a channel by making use of the energy and continuity equations. Remember that a similar problem occurred in pipes when a venturi meter was inserted to measure discharge (see Section 4.10). The equations of energy and continuity were used to work out the pressure and velocity changes as a result of sudden changes in the pipe diameter (Figure 5.17a). The solution was straight forward because the area of flow was fixed by the pipe diameter and so only the velocity needed to be calculated. But for a channel the flow area is not fixed. It is open to the atmosphere and so it can flow at many different depths (Figure 5.17b). So both the flow area and the velocity are unknown. If the energy and continuity equations are applied to this problem the result is a cubic equation which means there are three possible answers for the downstream depth

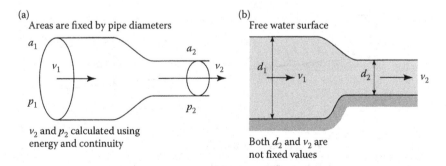

Figure 5.17 Predicting changes in depth and velocity in a channel is not straight forward. (a) Pipe flow. (b) Channel flow.

and velocity. One answer is negative and this can be dismissed immediately as impracticable. But the two remaining answers are both possibilities, but which one is the right one?

To help solve this problem, Boris Bakhmateff (1880–1951) introduced a very helpful concept which he called *specific energy* (E). Simply stated: *Specific energy is the energy measured from the bed of a channel.*

Writing this as an equation in terms of unit discharge (q)

$$E = d + \frac{q^2}{2gd^2}.$$

It is important at this point to draw a clear distinction between total energy and specific energy. They have similar components but are measured from different datums (Figure 5.18a).

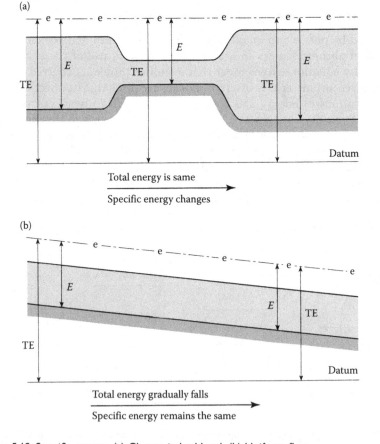

Figure 5.18 Specific energy. (a) Change in bed level. (b) Uniform flow.

Total energy is measured from some fixed datum and its value can only reduce as energy is lost through friction. When there is a change in the bed level of a channel (e.g. when water flows over a weir), there are also changes in the energy components but the total energy remains the same.

Specific energy is measured from the bed of a channel and so when the bed level changes, the specific energy also changes. It also means that specific energy can rise as well as fall depending on what is happening to the channel bed. When the flow moves from the channel over a weir, the specific energy falls and when it comes off the weir it rises again.

Uniform flow highlights the difference between total and specific energy (Figure 5.18b). Total energy falls gradually as energy is lost through friction. But specific energy remains constant along the channel because there are no changes in depth and velocity.

The physical significance of specific energy is not so obvious and many engineers still struggle with it. However, it is a very easy and very practical concept to use. Rather than be too concerned about what it means, it is better to think of it as a simple mechanism for solving a problem. It is like a key for opening a lock. You do not need to know how the lock works in order to use it. You just put the key in and turn it. In the same way, specific energy unlocks the problem of quantifying the effects that abrupt changes in a channel have on depth and velocity. It also helps to establish whether the flow is subcritical or supercritical, and it unlocks the problem of how to measure discharges in channels (see Section 6.5). So on the whole, it is a pretty effective key.

The best way to show how this works is with an example. A channel is carrying a discharge per unit width q. From the specific energy equation it is possible to calculate a range of values for specific energy (E) by putting in different values of d. When the values of d and E are plotted on a graph the result is the *specific energy diagram* (Figure 5.19a). There are two limbs to the diagram and this shows the two possible solutions for depth for any given value of E. These are called the *alternate depths* and are the sub and supercritical depths described earlier. The upper limb of the curve describes subcritical flow and the lower limb describes supercritical flow. The changeover between the two types of flow occurs at the point C on the graph and this is the *critical point*. It is the only place on the diagram where there is only one depth of flow for a given value of E and not two.

So the specific energy diagram clearly defines subcritical and supercritical flows by defining the critical depth. It also confirms the earlier descriptions of the effects on depth and velocity of sudden changes in a channel. Take any point E_1 on the subcritical part of the curve and look what happens when the value of E changes as a result of raising or lowering the channel bed. A step up on the bed reduces E and the graph shows that d also decreases (remember E is measured from the bed of the channel). A step down on the bed increases E and the graph shows that d increases. A similar example can be applied to the supercritical part of the diagram to show the effects of changing E on the depth of flow. In this case we see the opposite effect. A step up reduces E and increases d.

So the depth and velocity change, but by how much? This is where the specific energy diagram becomes very useful for quantifying these changes. To see how this is done, consider what happens when there is a step up of height h on the bed of a channel from point 1 to point 2 (Figure 5.19b). This reduces the specific energy E_2 by an amount h. So

$$E_1 - E_2 = h,$$

and

$$E_2 = E_1 - h.$$

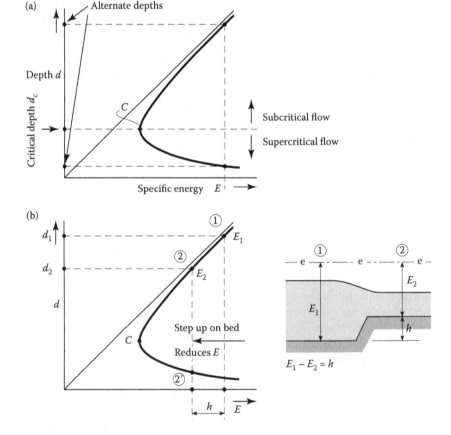

Figure 5.19 Specific energy diagram. (a) Alternate depths and critical point. (b) The effect on depths when there is a step in the channel bed.

E_2 on the curve can be found by subtracting h from the value of E_1 and this represents the condition on the step from which it is possible to determine the depth d_2 and velocity v_2. The same logic can also be applied to supercritical flow. An example in Box 5.5 shows how this is done in practice for both subcritical and supercritical flows.

One question often asked is: Why does the flow stay subcritical when it moves from point 1 to point 2? Why does it not go to the point 2′, which has the same value of specific energy, and so become supercritical? The answer comes from the specific energy diagram. Remember the diagram is for a given value of discharge. So suddenly 'leaping' across the diagram from 2 to 2′ would mean a change in discharge and this is not possible. The only way to get from 2 to 2′ is to go along the specific energy curve and through the critical point. This can only be done by increasing the value of h (i.e. raising the step even further) so that E_2 gets smaller until it reaches the critical point. At this point the flow can go supercritical. As h in this case is not high enough to create critical conditions, the flow moves from point 1 to point 2 and stays subcritical.

The same argument can be applied to changes in supercritical flow also. The flow can only go subcritical by going around the specific energy curve and through the critical depth.

To summarise some points about specific energy:

• Specific energy is used to quantify the changes in depth and velocity in a channel as a result of sudden changes in the size and shape of a channel. It also helps to determine which of the two possible answers for depth and velocity is the right one.
• Specific energy is different from total energy. It can increase as well as decrease. It depends on what happens to the bed of the channel. In contrast, total energy can only decrease as energy is lost through friction or sudden changes in flow.
• The specific energy diagram (Figure 5.19) is for one value of discharge. When the discharge changes a new diagram is needed. So for any channel there will be a whole family of specific energy diagrams representing a range of discharges.
• Specific energy is the principle on which many channel flow measuring devices, such as weirs and flumes, are based. They depend on changing the specific energy enough to make the flow go critical. At this point, there is only one value of depth for one value of specific energy and from this is possible to develop a formula for discharge (see Section 6.5). Such devices are sometimes referred to as *critical depth structures*.
• For uniform flow, the value of specific energy remains constant along the entire channel. This is because the depth and velocity are the same. In contrast, total energy gradually reduces as energy is lost along the channel.

5.7.3.1 Is the flow subcritical or supercritical?

The answer comes from calculating the normal depth of flow, using a formula such as Manning's, and then comparing it with the critical depth. If the depth is greater than the critical depth, the flow will be subcritical and if it is less it will be supercritical.

BOX 5.5   EXAMPLE: CALCULATING THE EFFECT ON DEPTH OF CHANGING THE CHANNEL BED LEVEL

A 0.3 m step is to be built in a rectangular channel carrying a discharge per unit width of 0.5 m^3/s (Figure 5.20). Calculate the effect of this change on the water level in both subcritical and supercritical flow conditions assuming the specific energy in the upstream channel $E_1 = 1.0$ m. Calculate the effect in the same channel of lowering a section of the channel bed by 0.3 m.

The first step is to calculate specific energy E for a range of depths for a discharge of 0.5 m^3/s:

Depth of flow (m)	Specific energy (m)
Subcritical flow curve	
0.4	0.480
0.5	0.551
0.6	0.635
0.7	0.726
0.8	0.820
0.9	0.916
1	1.013
1.2	1.209
1.4	1.407
1.6	1.605
Supercritical flow curve	
0.1	1.374
0.12	1.005
0.15	0.716
0.2	0.519
0.3	0.442
0.25	0.454
0.35	0.454

Plot the results on a graph (Figure 5.20a).

From the graph, the alternate depths for $E_1 = 1$ m are $d_1 = 0.99$ m (subcritical) and 0.12 m (supercritical). Consider the subcritical flow case first:

(a)

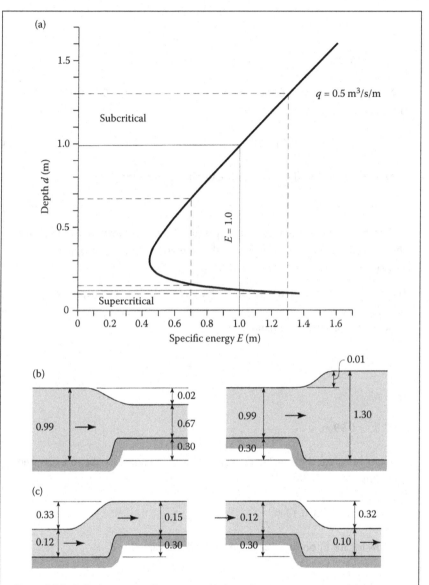

Figure 5.20 Calculating specific energy. (a) Specific energy diagram. (b) For sub-critical flow. (c) For supercritical flow.

For subcritical flow, calculate the effect on water level of a 0.3 m high step-up on the bed of the channel:

Specific energy in the channel is

$$E_1 = 1.0 \text{ m,}$$

and

$d_1 = 0.99$ m.

Next calculate the specific energy on the step which is 0.3 m higher than the channel bed

$E_2 = E_1 - 0.3 = 0.7$ m.

Now locate E_2 on the graph and read off the value for d_2

$d_2 = 0.67$ m.

Depth d_2 is measured from the step and so the water level is 0.97 m above the original channel bed. This means that the water level has dropped 0.02 m as a result of raising the channel bed.

For subcritical flow, calculate the effect on water level of a 0.3 m step-down on the bed of the channel:
Specific energy in the channel is

$E_1 = 1.0$ m,

and

$d_1 = 0.99$ m.

Next calculate the specific energy where the channel has been excavated by 0.3 m

$E_2 = E_1 + 0.3 = 1.3$ m.

Now locate E_2 on the graph and read off the value for d_2

$d_2 = 1.30$ m.

Depth d_2 is measured from the bed of the excavated section and so the water level is now 1.30 m above the channel bed. This means that the water level rises by 0.01 m as a result of the step down in the channel bed.

For supercritical flow, calculate the effect on water level of a 0.3 m step-up on the bed of the channel:

Specific energy in the channel is

$E_1 = 1.0 \text{ m},$

and

$d_1 = 0.12 \text{ m}.$

Next calculate the specific energy on the step which is 0.3 m higher than the channel bed

$E_2 = E_1 - 0.3 = 0.7 \text{ m}.$

Now locate E_2 on the graph and read off the value for d_2

$d_2 = 0.15 \text{ m}.$

Now depth d_2 is measured from the top of the raised bed and so the water level is now 0.45 m above the original channel bed. This means that the water level rises 0.33 m as a result of raising the channel bed.

For supercritical flow, calculate the effect on water level of a 0.3 m step-down on the bed of the channel:
Specific energy in the channel is

$E_1 = 1.0 \text{ m},$

and

$d_1 = 0.12 \text{ m}.$

Next calculate the specific energy where the channel has been excavated by 0.3 m

$E_2 = E_1 + 0.3 = 1.3 \text{ m}.$

Now locate E_2 on the graph and read off the value for d_2

$d_2 = 0.10 \text{ m}.$

Now depth d_2 is measured from the bed of the excavated section and so the water level is now 0.20 m below the original channel bed. This means that the water level drops 0.32 m as a result of the step down in the channel bed.

5.7.4 Critical depth

The critical depth d_c can be determined directly from the specific energy diagram, but it can be difficult to locate its exact position because of the rounded shape of the curve close to the critical point. To overcome this problem, it can be calculated using a formula derived from the specific energy equation

$$d_c = \sqrt[3]{\frac{q^2}{g}}.$$

This formula shows that the critical depth is influenced only by the discharge per unit width q. It has nothing to do with the slope or the normal depth. For the mathematically minded, a derivation of this is given in Box 5.6 and an example in Box 5.7.

BOX 5.6   DERIVATION: CRITICAL DEPTH EQUATION

Derive a formula for the critical depth and its location on the specific energy diagram (Figure 5.19a).

Use the specific energy equation

$$E = d + \frac{q^2}{2gd^2}.$$

At the critical point, the specific energy E is at its lowest value and so the gradient of the curve (dE/dd) is equal to zero. The equation for the gradient can be found using calculus and differentiating the above equation for the curve. If you are not familiar with calculus you will have to accept this step as given

$$\frac{dE}{dd} = 1 - \frac{q^2}{gd^3} = 0.$$

Depth d now becomes the critical depth d_c and so

$$\frac{q^2}{gd_c^3} = 1.$$

Rearrange this for d_c:

$$d_c^3 = \frac{q^2}{g},$$

$$d_c = \sqrt[3]{\frac{q^2}{g}}.$$

So the critical depth d_c depends only on the discharge per unit width q.

To calculate the specific energy at the critical point, first write down the specific energy equation for critical conditions

$$E_c = d_c + \frac{q^2}{2gd_c^2}.$$

But at the critical depth (see equation earlier in this proof)

$$\frac{q^2}{gd_c^3} = 1.$$

Dividing both sides by 2 and multiplying by d_c

$$\frac{q^2}{2gd_c^2} = \frac{d_c}{2}.$$

The left hand side is now equivalent to the kinetic energy term, so put this into the specific energy equation

$$E_c = d_c + \frac{d_c}{2} = \frac{3d_c}{2}.$$

So for any given discharge, the critical depth can be calculated and also the specific energy at the critical point. These two values exactly locate the critical point on the specific energy diagram.

Note also from the above analysis that when

$$\frac{q^2}{gd_c^3} = 1,$$

$$v_c^2 = gd_c,$$

and so

$$v_c = \sqrt{gd_c}.$$

This is the equation for the celerity of surface water waves and it shows that at the critical point the water velocity v_c equals the wave celerity $\sqrt{gd_c}$. Remember that the velocity of waves across water following a disturbance is often used to determine whether flow is subcritical or supercritical.

BOX 5.7   EXAMPLE: CALCULATING THE STEP UP REQUIRED TO ACHIEVE CRITICAL FLOW

Using information in the previous example (Box 5.5), calculate the critical depth and the step up in the bed level required to ensure that the flow will reach the critical depth. Assume the initial specific energy $E_1 = 1.0$ m.

First calculate the critical depth

$$d_c = \sqrt[3]{\frac{q^2}{g}},$$

$$d_c = \sqrt[3]{\frac{0.5^2}{9.81}} = 0.29 \text{ m}.$$

Now calculate the specific energy on the step up

$$E_2 = d_2 + \frac{q^2}{2gd_2^2}.$$

When this is critical flow

$$d_2 = d_c,$$

and so

$$E_2 = 0.4 + \frac{0.5^2}{2 \times 9.81 \times 0.29^2} = 0.44 \text{ m},$$

but

$$E_1 - E_2 = h,$$

that is, the change in specific energy is a direct result of the height of the step up h and so

$$h = 1.0 - 0.44 = 0.56 \text{ m}.$$

The bed level of the channel must be raised by 0.56 m to ensure that the flow goes critical.

5.7.5 Critical flow

Although critical flow is important for measuring purposes, it is a flow condition best avoided in uniform flow. There is no problem when flow goes quickly through the critical point on its way from subcritical to supercritical or from supercritical to subcritical, but there are problems when the normal flow depth is near to the critical depth. This is a very unstable condition as the flow tends to oscillate from sub to super and back to subcritical again resulting in surface waves which can travel for many kilometres eroding and damaging channel banks. The explanation for the instability is in the shape of the specific energy diagram close to the critical point (Figure 5.19a). Small changes in specific energy E, possibly as a result of small channel irregularities, can cause large changes in depth as the flow oscillates between sub- and super-critical flow. As the flow *hunts* back and forth it sets up waves. So although critical flow is very useful in some respects, it can cause serious problems in others.

5.7.6 Flow transitions

Changes in a channel which result in changes in flow from subcritical to supercritical and vice versa are referred to as *transitions*. The following are examples of some common transitions.

5.7.6.1 Subcritical to supercritical flow

When flow goes from subcritical to supercritical, it does so smoothly. In Figure 5.21a, the channel slope is increased which changes the flow from sub to supercritical. The water surface curves rapidly but smoothly as the flow goes through the critical point. There is no energy loss as the flow is contracting. Notice how the critical depth is shown as c—c—c so that the two types of flow are clearly distinguishable. Remember, the critical depth is the same in both sections of the channel because it depends only on the discharge and not on slope.

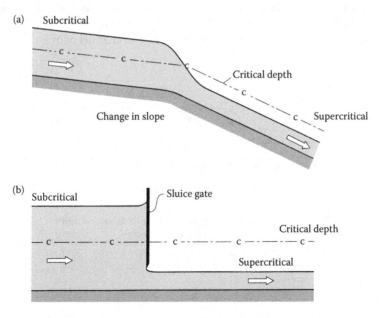

Figure 5.21 Flow transitions – sub to supercritical. (a) Change in channel slope. (b) Flow under a sluice gate.

Another example of this type of transition is the sluice gate (Figure 5.21b). In this case a gate is used to force the change in flow. Again the transition occurs smoothly with little energy loss.

5.7.6.2 Supercritical to subcritical flow (hydraulic jump)

The change from supercritical to subcritical is not so smooth. In fact, a vigorous turbulent mixing action occurs as the flow jumps abruptly from supercritical to subcritical flow (Figure 5.22). It is aptly called a *hydraulic jump* and as the flow expands there is a significant loss of energy due to the turbulence.

Hydraulic jumps are very useful for many purposes:

- Getting rid of unwanted energy, such as at the base of dam spillways.
- Mixing chemicals in water. The vigorous turbulence ensures that any added chemical is thoroughly dispersed throughout the flow.
- Converting supercritical flow downstream of hydraulic structures into subcritical flow to avoid erosion damage in unprotected channels.

Hydraulic jumps are usually described by their strength and the Froude No. of the supercritical flow. A *strong jump* is the most desirable. It is very vigorous, has a high Froude No., well above one and the turbulent mixing is confined to a short length of channel. A *weak jump,* on the other hand, is not so violent. It has a low Froude No. approaching one, which means the depth of flow is close to the critical depth. This kind of jump is not confined and appears as waves which can travel downstream for many kilometres. This is undesirable because the waves can do a great deal of damage to channel banks.

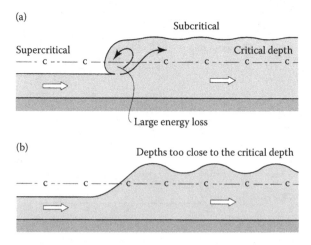

Figure 5.22 Flow transitions – super to subcritical. (a) Strong hydraulic jump. (b) Weak hydraulic jump.

5.7.6.3 Creating a hydraulic jump

There are two conditions required to form a hydraulic jump:

- Flow upstream of the jump must be supercritical, that is, the Froude No. must be greater than 1.
- Downstream flow must be subcritical and deep enough for the jump to form.

Once there is supercritical flow in a channel, it is the downstream depth of flow that determines if a jump will occur. To create a jump, the downstream depth must be just right. If the depth is too shallow, a jump will not form and the supercritical will continue down the channel (Figure 5.23). Conversely if the flow is too deep, the jump will move upstream and if it reaches a sluice gate, it may drown it out. This can be a problem as the high-speed supercritical flow is not dispersed as it would be in a full jump and it can cause erosion downstream (see Section 6.2 for drowned flow from a sluice gate).

When the upstream depth and velocity are known, it is possible to calculate the downstream depth and velocity which will create a jump by using the momentum equation. The energy equation cannot be used at this stage because of the large and unknown energy loss at a jump.

The formula is derived from the momentum equation which links the two depths of flow d_1 and d_2

$$\frac{d_2}{d_1} = \frac{1}{2}\left(\sqrt{1 + 8F_1^2} - 1\right),$$

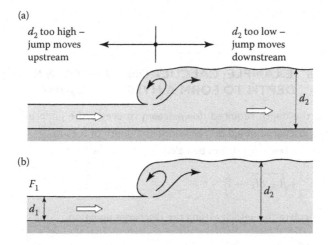

Figure 5.23 Forming a hydraulic jump. (a) Jump moves when downstream depth is not right. (b) Calculating the right downstream depth.

where d_1 is upstream depth, d_2 is downstream depth and F_1 is upstream Froude No. (See the example for calculating the downstream depth in Box 5.8.)

Most hydraulic jump problems involve calculating the downstream depth d_2 in order to determine the conditions under which a jump will occur. But determining a value for d_2 is fine in theory, but in practice the formation of a jump is very sensitive to the downstream depth. This means that when the downstream depth is a little more or less than the calculated value, the jump does not stabilise at the selected location but moves up and down the channel *hunting* for the right depth of flow. In practice, it is very difficult to control water depths with great accuracy and so some method is needed to remove the sensitivity of the jump to downstream water level and stabilise it. This is the job of a *stilling basin*, which is a concrete apron located downstream of a weir or sluice gate. Its primary job is to dissipate unwanted energy to protect the downstream channel and it does this by making sure the hydraulic jump forms on the concrete slab even though the downstream water level may vary considerably. This is discussed further in Section 6.10.

5.7.6.4 Calculating energy losses

Once the downstream depth and velocity have been calculated using the momentum equation, the loss of energy at a jump can be determined using the total energy equation as follows

Total energy upstream = total energy downstream + energy loss,

that is,

$$d_1 + \frac{v_1^2}{2g} + z_1 = d_2 + \frac{v_2^2}{2g} + z_2 + \text{ losses.}$$

BOX 5.8   EXAMPLE: CALCULATING THE DOWNSTREAM DEPTH TO FORM A HYDRAULIC JUMP

Calculate the depth required downstream to create the jump in a channel carrying a discharge of 0.8 m³/s/m width at a depth of 0.25 m.

The downstream depth can be calculated using the formula

$$\frac{d_2}{d_1} = \frac{1}{2}\left(\sqrt{1 + 8F_1^2} - 1\right).$$

First calculate the velocity using the discharge equation

$$q = v_1 d_1,$$

$$v_1 = \frac{0.8}{0.25} = 3.2 \text{ m/s.}$$

Next calculate the Froude No.

$$F_1 = \frac{v_1}{\sqrt{gd_1}},$$

$$= \frac{3.2}{\sqrt{9.81 \times 0.25}} = 2.04.$$

Substitute the values into the above formula

$$\frac{d_2}{0.25} = \frac{1}{2}\left(\sqrt{1 + (8 \times 2.04^2)} - 1\right),$$

$$d_2 = 0.25 \times 2.43 = 0.61 \text{ m.}$$

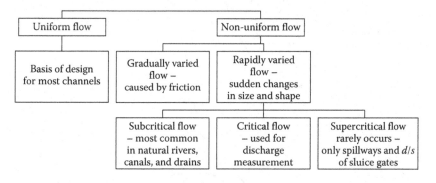

Figure 5.24 Summary of various types of channel flow.

Figure 5.24 summarises all the various types of flow that can occur in channels.

5.8 SECONDARY FLOWS

One interesting aspect of channel flow (which all anglers know about) is *secondary flows*. These are small but important currents that occur in flowing water and explain many important phenomena.

In ideal channels, the velocity is assumed to be the same across the entire channel. But in real channels, the velocity is usually much higher in the middle than at the sides and bed. Figure 5.25 shows the velocity profile in a typical channel and in cross-section, isotachs (lines of equal velocity) have been drawn. The changes in velocity across the channel cause small changes in pressure

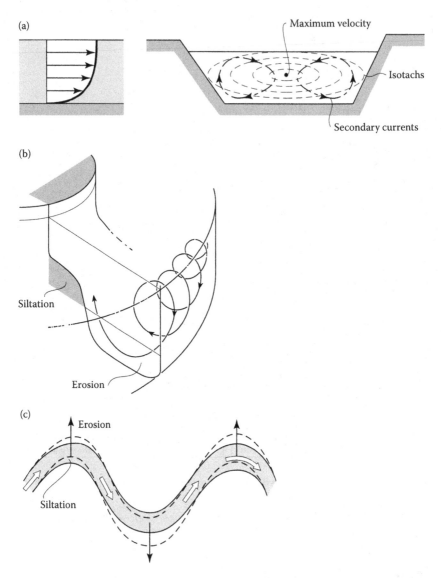

Figure 5.25 Secondary flows in channels. (a) Velocity profiles. (b) Flow around a river bend. (c) River meander pattern.

(remember the energy equation) and these are responsible for setting up cross currents, which flow from the sides of the channel to the centre. As the water does not pile up in the middle of a channel, there must be an equivalent flow from the centre to the edge. These circulating cross flows are called *secondary flows*. In very wide channels, several currents can be set up in this way. So water does not just flow straight down a channel, it flows along a spiral path.

5.8.1 Channel bends

An important secondary flow occurs at channel bends (Figure 5.25b). This is created by a combination of the difference in velocity between the surface and the bed and the centrifugal forces as the water moves around the bend. A secondary current is set up which moves across the bed from the outside of the bend to the inside and across the surface from the inside to the outside. The secondary flow can erode loose material on the outside of a bend and carry it across a river bed and deposit it on the inside of the bend. This is contrary to the common belief that sediment on the bed of the river is thrown to the outside of a bend by the strong centrifugal forces.

One consequence of this in natural erodible river channels is the process known as *meandering* (Figure 5.25c). Very few natural rivers are straight. They tend to form a snake-like pattern of curves which are called *meanders*. The outside of bends are progressively scoured and the inside silts up causing the river cross section to change shape. The continual erosion gradually alters the course of the river. Sometimes the meanders become so acute that parts of the river are eventually cut off and form what are called *ox-bow lakes*.

Note that it is not a good idea to go swimming on the outside bend of a river. The downward current can be very strong and pull the unwary swimmer down into the mud on the river bed!

5.8.2 Siting river off-takes

A common engineering problem is to select a site for abstracting water from a river for domestic use or irrigation. This may be a pump or some gated structure. The best location is on the outside of a river bend so that it will be free from silting (Figure 5.26). If located on the inside of a bend, it would be continually silting up as a result of the actions of the secondary flows. The outside of a bend may need protecting with stone pitching to stop any further erosion which might destroy the off-take.

5.8.3 Bridge piers

Scouring around bridge piers can be a serious problem. This is the result of a secondary flow set up by the stagnation pressure on the nose of a pier (Figure 5.27). The rise in water level as the flow is stopped causes a secondary downward current towards the bed. This is pushed around the pier by the main flow into a spiral current which can cause severe scouring both in front and around the sides of piers. Heavy stone protection can reduce the problem but a study of the secondary flows has shown that the construction of low walls upstream can also help by upsetting the pattern of the destructive secondary currents.

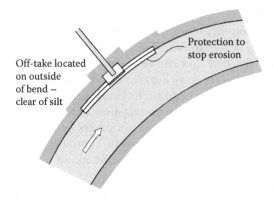

Figure 5.26 Siting river off-takes.

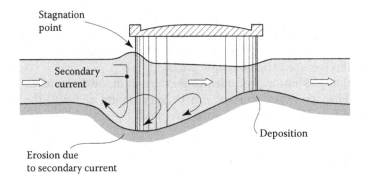

Figure 5.27 Flow around bridge piers.

5.8.4 Vortices at sluice gates

Vortices can develop upstream of sluice gates and may be so strong that they extend from the water surface right underneath the gate drawing air down into its core (Figure 5.28). They can become so severe that they reduce the

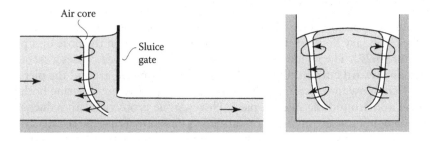

Figure 5.28 Vortices at sluice gates.

flow through the gate. They look like small tornadoes and are the result of secondary currents. They form when the surface flow reaches a sluice gate and the water stops, that is, it stagnates. But the flow in the middle of the stream was moving faster than near the sides and so the rise in water level due to the stagnation is higher in the middle than at the sides. This causes a secondary flow from the middle of the stream to the sides and sets up two circulating vortices.

5.8.5 Tea cups

There is a great deal more going on in a tea cup than brewing the tea (Figure 5.29). First make sure there is no milk in the tea to obscure the view, then notice how the tea leaves at the bottom of the cup move in towards the centre when you stir it. The stirring action sets up a vortex which is similar to the flow round a bend in a river. The water surface drops in the centre of the flow causing a pressure difference which results in a downward current on the outside of the cup which moves across the bottom and up the middle. Any tea-leaves lying on the bottom of the cup are swept along a spiral path towards the centre of the cup.

5.9 SEDIMENT TRANSPORT

Sediment movement in channels is usually referred to as *sediment transport* and it is not a subject normally covered in texts on basic hydraulics. But a study of channels would not be complete without mentioning it. Most

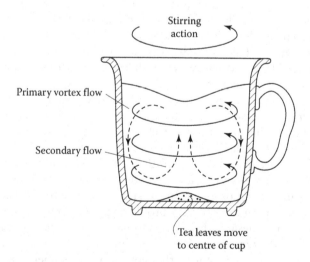

Figure 5.29 Stirring the tea.

channels are either naturally occurring or excavated in the natural soil and so are prone to scouring and silting. Very few have the luxury of a lining to protect them against erosion. Natural rivers often carry silt and sand washed from their catchments; some even carry large boulders in their upper reaches. Man-made canals not only have to resist erosion but need to avoid becoming blocked with silt and sand. Indeed the sediment is often described as the sediment load, indicating the burden that channels must carry.

5.9.1 Channels carrying sediment

Channels carrying sediment are designed in a different way to those carrying clear water. Normally, Manning or Chezy formulae are used for clear water but they are not well suited to sediment laden water. There is no simple accepted theory which provides a thorough understanding of sediment movement on which to base channel design. However, when engineers meet a problem for which there is no acceptable theory, they do not wait for the scientists to find one. They look for a way round the problem and develop alternative design methods. This is what happened in India and Pakistan at the turn of last century when British engineers where faced with designing and operating large canals which carried both water and silt from the Indus and Ganges rivers for the irrigation of vast tracts of land (see Section 8.5).

They observed and measured the hydraulic parameters of existing canals that seemed to have worked well under similar conditions over a long period of time. From these data they developed equations which linked together sediment, velocity, hydraulic radius, slope and width and used them to design new canals. They came to be known as the *regime equations*. The word regime described the conditions in the canals where, over a period of time, the canals neither silted nor scoured and all the silt that entered the canals at the head of the systems was transported through to the fields. In fact an added benefit of this was the silt brought with it natural fertilisers for the farmers. But this is not claimed to be a perfect solution. Often the canals silted up during heavy floods and often they would scour when canal velocities where excessive. But on balance over a period of time (which may be several years) the canals did not change much and were said to be *in regime*. In spite of all the research over the past century, engineers still use these equations because they are still the best available today.

The lack of progress in our understanding of sediment transport comes from the nature of the problem. In other aspects of hydraulics, there is a reasonably clear path to solving a problem. In channels, one single force dominates channel design, the gravity force. All other forces, such as viscosity, can be safely ignored without causing much error. But in sediment transport, there are several forces that control what happens and no single one dominates. Gravity influences both water and sediment, but viscosity and density also appear to be important. This makes any fundamental analysis of the problem complex.

Interestingly, the physics of sediment movement in air, such as windblown sand dunes, is well understood which was set down by Ralph Bagnold (1896–1990). As a British army officer, he spent much of his time in the desert studying the movement of sand dunes and became one of the world's leading experts. He was also the founder of the British Army's Long Range Desert Group which was the forerunner of today's SAS. Unfortunately, the physics of sediment movement in water still evades our full understanding.

5.9.2 Threshold of movement

At the heart of the sediment transport problem is how to determine when sediment begins to move: the *threshold of movement*. Some investigators, particularly engineers, have tried to link this with average stream velocity because it is simple to measure and it is obviously linked to the erosive power of a channel. But research has shown that it is not as simple as this. Look at the particles of sand on the bed of a channel (Figure 5.30). Is it the average velocity that is important or the local velocity and turbulence close to the sand grains? Some grains are more exposed than others. Do they move first? Does grain size and density matter? Do the rounded grains move more easily than the angular ones? Is there a Bernoulli (energy change) lifting effect when flow moves around exposed grains that encourage them to move? How important is the apparent loss in weight of sand grains when they are submerged in water? There are lots of questions and no clear answers. But what is clear is that many factors influence the threshold of movement and this has made it a very difficult subject to study from an analytical point of view.

There is however some good news which has come from experimental work. It is in fact very easy to observe the threshold of movement and to say when it has been reached. Imagine a channel with a sandy bed with water flowing over it. When the flow is increased, the sand will at some point

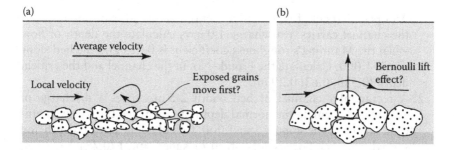

Figure 5.30 Threshold of movement. (a) The mystery of how particles begin to move. (b) Is there a 'lift' effect from small changes in energy.

begin to move. But the interesting point is that this begins quite suddenly and not gradually as might be expected. When the threshold is reached, the whole channel bed comes alive suddenly as all the sand begins to move at the same time. So if several observers, watching the channel, are asked to say when they think the threshold has been reached, they will have no problem in agreeing the point at which it occurs. What they will not be able to say for certain is what caused it.

Because of this clear observation of the threshold, much of the progress had been made by scientists using experimental methods. Shields in 1936 successfully established the conditions for the threshold of movement on an experimental basis for a wide range of sediments and these are the data that are still used today for designing channels to avoid erosion. It provides a much sounder basis for design than simply using some limiting velocity.

5.9.3 Calculating sediment transport

Working out the amount of sediment being transported once it begins to move is fraught with difficulties. The reason for this is that once movement begins the amount of sediment on the move is very sensitive to small changes in the factors which caused the movement in the first place. This means that small changes in what could be called the erosive power of a channel can result in very large changes in sediment transport. Even if it was possible to calculate such changes, which some experimenters have tried to do, it is even more difficult to verify this by measuring sediment transport in the laboratory and almost impossible to measure it with any accuracy in the field. For these reasons, it seems unlikely that there will be any significant improvements in the predictions of sediment transport and that engineers will have to rely on the regime equations for some time to come.

5.10 SOME EXAMPLES TO TEST YOUR UNDERSTANDING

1. An open channel of rectangular section has a bed width of 1.0 m. If the channel carries a discharge 1.0 m³/s calculate the depth of flow when the Manning's roughness coefficient is 0.015 and the bed slope is 1 in 1,000. Calculate the Froude No. in the channel and the critical depth (0.5 m; 0.63; 0.29 m).
2. A rectangular channel of bed width 2.5 m carries a discharge of 1.75 m³/s. Calculate the normal depth of flow when the Chezy coefficient is 60 and the slope is 1 in 2,000. Calculate the critical depth and say whether the flow is subcritical or supercritical (0.75 m; 0.37 m; flow is subcritical).

3. A trapezoidal channel is to be designed and constructed in a sandy loam with a longitudinal slope of 1 in 5,000 to carry a discharge of 2.3 m³/s. Calculate suitable dimensions for the depth and bed width assuming Manning's n is 0.022 and the side slope is 1 in 2 ($d = 0.98$ m; $b = 2.94$ m; when $b = 3d$).

4. A trapezoidal channel carrying a discharge of 0.75 m³/s is to be lined with concrete to avoid seepage problems. Calculate the channels dimensions which will minimise the amount of concrete when Manning's n for concrete is 0.015 and the channel slope is 1 in 1,250 (for a hexagonal channel $d = 0.37$ m; $b = 0.8$ m. Note that other answers are possible depending on choice of side slope).

5. A hydraulic jump occurs in a rectangular channel 2.3 m wide when the discharge is 1.5 m³/s. If the upstream depth is 0.25 m calculate the upstream Froude No., the depth of flow downstream of the jump and the energy loss in the jump (2.78; 0.87 m; 0.3 m).

Chapter 6

Hydraulic structures
for channels

6.1 INTRODUCTION

Hydraulic structures in channels have three main functions:

- Measuring and controlling discharge
- Controlling water levels
- Dissipating unwanted energy

Measuring and controlling discharge in channels is perhaps one of the more obvious uses of hydraulic structures (Figure 6.1). Large irrigation networks require structures at each canal junction to measure and control discharge so that there is a fair and equitable distribution of water. It is not enough just to construct a canal junction and hope that the flow will divide itself properly between the two. Natural rivers too need regular flow measurement so that engineers can make sure there is an adequate supply to meet the growing demands for domestic and industrial uses as well as maintaining base flows for environmental purposes. It is important to ensure that minimum base flows are maintained in dry summer periods to protect fish stocks and environmentally sensitive wetlands. Flood flows are also measured so that adequate precautions can be taken to avoid or control flooding, particularly in urban areas where damage can be very costly.

The need to control water levels in channels may not be so obvious. On irrigation schemes, water level control is just as important as discharge control. The canals are built up higher than the surrounding ground level so there is enough energy for water to flow by gravity through pipes from the channels into the farms (Figure 6.2a). The water level (known as the *command*) must be carefully controlled if each farm is to receive the right discharge. But all too often the water level drops because of low flows or seepage losses. This reduces the discharge through the pipes making it difficult for farmers to irrigate properly. To avoid this problem, hydraulic structures are built across the canals to raise the water levels back to their command levels. Such structures are called *cross regulators*.

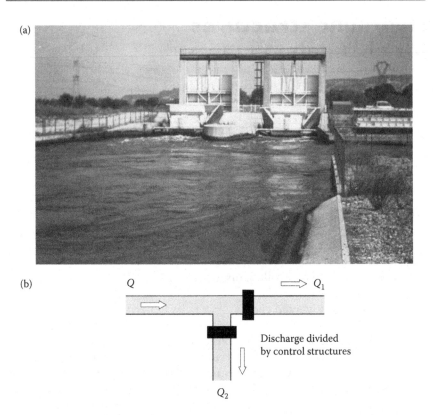

Figure 6.1 Measuring and controlling discharge. (a) Discharge regulator in a canal system in France. (b) Control structures divide the flow.

Another example is to control river water level close to a natural wetland site which is valued for its bird population or its special plants (Figure 6.2b). Water enters the site either by flooding from the river as it over tops its banks or through seepage from its bed and sides. Either way the wetland is very dependent on the water level in the river as well as its flow to avoid it drying out and causing irreparable damage to flora and fauna. Fluctuations in water level can be avoided by building a structure across the river to hold the water at the desired level throughout the year even though the discharge may vary significantly.

Hydraulic structures are also very useful for getting rid of unwanted energy. When water flows down dam spillways, it can reach speeds of 60 km/h and more and is capable of doing a lot of damage. Hydraulic structures are used to stop such high-speed flows and dissipate the kinetic energy by creating hydraulic jumps.

Some hydraulic structures only carry out one of the functions described above, whilst others perform all three functions at the same time. So a hydraulic structure may be used for discharge measurement, and at the

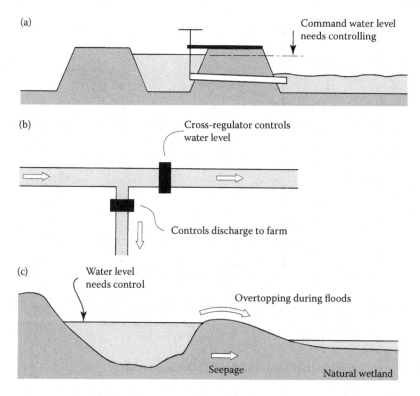

Figure 6.2 Controlling water levels. (a) Command water level. (b) Cross regulator to control water level. (c) Controlling wetland flooding.

same time, it may be performing a water level control function and dissipating unwanted energy.

From the point of view of measuring discharge and controlling water levels, there are only two types of structure. Some structures allow water to flow through them and these are called *orifice structures*. Others allow water to flow over them and these are called *weirs or flumes*. Hydraulically, they behave in quite different ways and so each has certain applications for which they are best suited. The energy dissipating function can be attached to both of these structure types.

6.2 ORIFICE STRUCTURES

The principles of orifice flow were described earlier in Section 3.6.3. In practice, orifice structures have fixed or moveable gates rather than just a simple opening. The sluice gate described in Section 5.4.3 is a good example of this type of structure. The flow under the gate is very similar to orifice flow but not quite (Figure 6.3a). First, the flow contracts only on its upper surface

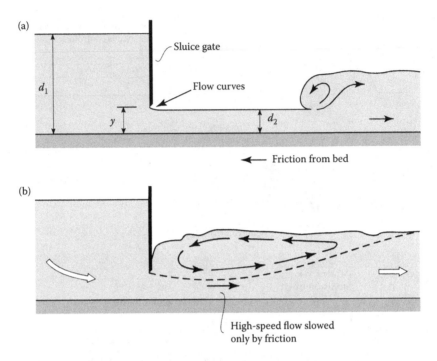

Figure 6.3 Orifice structures. (a) Free flow. (b) Drowned flow.

as it goes under the gate, and second, there is additional friction from the bed of the channel. So to find a formula for discharge for this structure, the orifice formula is a good starting point but it needs modifying.

The formula for discharge from an orifice is

$$Q = a\sqrt{2gh}.$$

Modifying this for a sluice gate

$$Q = C_d a\sqrt{2gd_1},$$

where Q is discharge (m³/s), a is area of gate opening (m²), C_d is a coefficient of discharge and d_1 is the water depth upstream of the orifice (m).

This looks to be a simple formula. It is an attempt to relate the discharge to the area of the gate opening a and the upstream water depth d_1 because both are easy to measure. However, the relationship is not so simple and as a result C_d is not a simple coefficient of discharge as defined earlier in Section 4.10. It takes account of the contraction of the flow under the gate and it also allows for energy losses and the effects of the size of gate opening and the head on the gate. So C_d is not a simple constant number like the coefficient of contraction C_c. Usually, the manufacturer of hydraulic gates will supply suitable values of C_d so that the 'simple' discharge formula can be used.

6.2.1 Free and drowned flow

The sluice gate example shows the flow freely passing under the gate with a hydraulic jump downstream. The downstream depth d_2 has no effect on the upstream depth d_1. This is referred to as free flow and the formula quoted above for calculating discharge is based on this condition.

In some circumstances, the jump can move upstream and drown out the gate and is referred to as drowned flow (Figure 6.3b). The flow downstream may look very turbulent and have the appearance of a jump, but inside the flow, the action is quite different. There is very little turbulent mixing taking place and the supercritical flow is shooting underneath the subcritical flow. This high-speed jet is not stopped quickly as it would in a jump, but slows down gradually over a long distance through the forces of friction on the channel bed. This flow can do a lot of damage to an unprotected channel even though the water surface may appear to be quite tranquil on the surface. Under drowned conditions, the formula for discharge must be modified to take account of the downstream water level which now has a direct influence on the upstream water level.

6.3 WEIRS AND FLUMES

Weirs and flumes are both overflow structures with very different characteristics to orifices. Many different types of weirs have been developed to suit a wide range of operating conditions, some comprise just a thin sheet of metal across a channel (*sharp-crested weirs*), whereas others are much more substantial (*solid weirs*). Both are based on the principle of changing the energy in a channel and using the energy and continuity equations to develop a formula for discharge based on depth (pressure) measurements upstream. But solid weirs rely on an energy change which is sufficient to make the flow go through the critical point. Because of this, they are sometimes called *critical-depth structures*. Sharp-crested weirs do not have this constraint – but they do have others.

Flumes are also critical-depth structures but rely on changing the energy by narrowing the channel width rather than raising the bed. Sometimes engineers combine weirs and flumes by both raising the bed and narrowing the width to achieve the desired energy change. It does not really matter which way critical flow is achieved so long as it occurs.

6.4 SHARP-CRESTED WEIRS

Sharp-crested weirs are used to measure relatively small discharges (Figure 6.4a). They comprise a thin sheet of metal such as brass or steel (sometimes wood can be used for temporary weirs) into which a specially

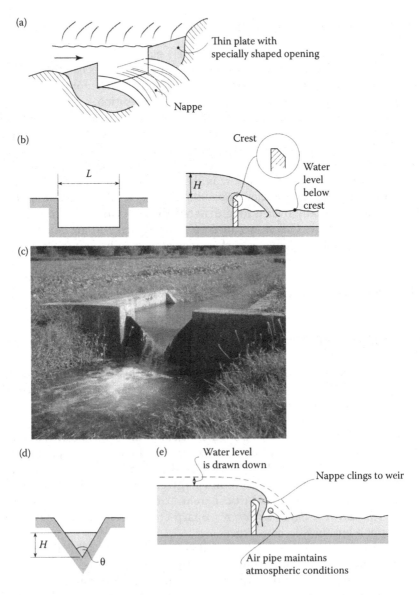

Figure 6.4 Sharp-crested weirs. (a) Sharp-crested weir. (b) Rectangular weir details including crest. (c) V-shaped weir. (d) V-shaped weir details. (e) Problem when nappe is not aerated.

shaped opening is cut. This must be accurately cut leaving a sharp edge with a bevel on the downstream side. When located in a channel, the thin sheet is sealed into the bed and sides so that all the water flows through the opening. By measuring the depth of water above the opening, known as the head on the weir, the discharge can be calculated using a formula derived from the energy equation. There is a unique relationship between the head on the weir and the discharge and one simple depth measurement determines the discharge.

6.4.1 Rectangular weirs

This weir has a rectangular opening (Figure 6.4b). Water flows through this and plunges downstream. The overflowing water is often called the *nappe*. The discharge is calculated using the formula

$$Q = \frac{2}{3} C_d L \sqrt{2g} H^{1.5},$$

where C_d is a coefficient of discharge, L is length of weir (m) and H is the head on the weir measured above the crest (m).

C_d allows for all the discrepancies between theory and practice.

6.4.2 Vee-notch weirs

This weir has a triangular-shaped notch and is ideally suited for measuring small discharges (Figure 6.4c and d). If a rectangular weir was used for low flows, the head would be very small and difficult to measure accurately. Using a vee weir, the small flow is concentrated in the bottom of the vee providing a reasonable head for measurement.

The discharge is calculated using the formula

$$Q = \frac{8}{15} C_d \sqrt{2g} \, Tan \frac{\theta}{2} H^{2.5},$$

where θ is the angle of the notch.

6.4.3 Some practical points

There are several conditions that must be met for these weirs to work properly. These are set out in detail in British Standard BS3680 (see Further Reading for details). The following are some of the key points:

- The water must fall clear of the weir plate into the downstream channel. Notice the bevelled edge on the crest facing downstream which creates the sharp edge and helps the water to spring clear.

If this did not happen, the flow clings to the downstream plate and draws down the flow reducing the head on the weir (Figure 6.4d). Using the above formula with the reduced value of head would clearly not give the right discharge.

- Flow over the weir must be open to the atmospheres so that the pressure around it is always atmospheric. Sometimes the falling water draws air from underneath the weir, and unless this air is replaced, a vacuum may form which causes the flow to cling to the downstream face (Figure 6.4d). This draws down the upstream water level reducing the head on the weir and giving a false value of discharge when it is put into the formula. To prevent this, air must be allowed to flow freely under the nappe.
- The weir crest must always be set above the downstream water level. This is the *free flow* condition for sharp-crested weirs. If the downstream level rises beyond the crest, it starts to raise the upstream level and so the weir becomes *drowned*. Another word that is used to describe this condition is *submerged flow*. The formula no longer works when the flow is drowned and so this situation must be avoided by careful setting of the weir crest level.
- The head H must be measured a few metres upstream of the weir to avoid the draw-down effect close to the weir.
- When deciding what size of weir to use, it is important to make sure there is a reasonable head so that it can be measured accurately. This implies that you need to have some idea of the discharge to be measured before you can select the right weir size to measure it. If the head is only say 4 mm, then 1 mm error in measuring it is a 25% error and will result in a significant error in the discharge. However, if the head is 100 mm then a 1 mm error in measuring the head is only a 1% error and so is not so significant.

Sharp-crested weirs can be very accurate discharge measuring devices, provided they are constructed carefully and properly installed. However, they can be easily damaged, in particular the sharp crest. If this becomes rounded or dented through impact with floating debris, the flow pattern over the weir changes and this reduces its accuracy. For this reason, they tend to be unsuited for long-term use in natural channels but well suited for temporary measurements in small channels, in places where they can be regularly maintained and for accurate flow measurement in laboratories.

6.5 SOLID WEIRS

These are much more robust than sharp-crested weirs and are used extensively for flow measurement and water level regulation in rivers and canals (Figure 6.5a).

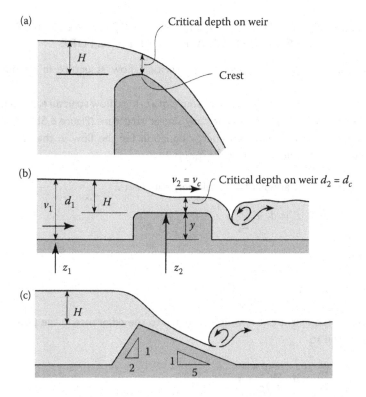

Figure 6.5 Solid weirs. (a) Solid weir. (b) Broad-crested weir. (c) Crump weir.

All solid weirs work on the principle that *the flow over the weir must go through the critical depth*. The idea of critical depth was discussed in Chapter 5, where the height of a weir, referred to earlier as a step in the channel bed, determines whether or not the flow goes critical. Once this happens, a formula for discharge can be developed using specific energy and the special conditions that occur at the critical point.

The formula links the channel discharge (Q) with the upstream water depth *measured above the weir crest* (H).

$$Q = CLH^{1.5},$$

where C is weir coefficient, L is length of the weir crest (m) and H is head on the weir measured from the crest (m).

For details about how this formula is developed see Box 6.1.

As there is some draw-down close to the weir, the head is usually measured a few metres upstream where the water level is unaffected by the weir.

BOX 6.1   DERIVATION: FORMULA FOR DISCHARGE OVER A CRITICAL FLOW STRUCTURE

Derive a formula for discharge for a critical flow structure in an open channel.

Use a broad-crested weir as an example of a critical flow structure, although the analysis would be the same for any similar structure (Figure 6.5b).

First write down the total energy equation for the flow in the channel (point 1) and the flow on the weir (point 2)

Total energy at point 1 = Total energy at point 2

$$d_1 + \frac{v_1^2}{2g} + z_1 = d_2 + \frac{v_2^2}{2g} + z_2.$$

Now

$$z_2 - z_1 = y.$$

This is the height of the weir. It is equal to the difference in the potential energy and so

$$d_1 + \frac{v_1^2}{2g} = d_2 + \frac{v_2^2}{2g} + y.$$

But at the critical depth

$$d_2 = d_c.$$

And $v_2 = v_c$. Substitute these into the energy equation

$$d_1 + \frac{v_1^2}{2g} = d_c + \frac{v_c^2}{2g} + y.$$

But also at the critical depth

$$\frac{v_c^2}{2g} = \frac{d_c}{2}.$$

(See derivation of the formula for critical depth in Box 5.6). Put this into the equation

$$d_1 + \frac{v_1^2}{2g} = d_c + \frac{d_c}{2} + y.$$

Rearrange this

$$d_1 + \frac{v_1^2}{2g} - y = \frac{3}{2}d_c.$$

But critical depth can be calculated from the formula (derivation in Box 5.6)

$$d_c = \sqrt[3]{\frac{q^2}{g}}.$$

Substituting for d_c

$$d_1 + \frac{v_1^2}{2g} - y = \frac{3}{2}\sqrt[3]{\frac{q^2}{g}}.$$

Now put

$$d_1 + \frac{v_1^2}{2g} - y = H.$$

Note that H is measured from the weir crest. Substitute this in the above energy equation

$$H = \frac{3}{2}\sqrt[3]{\frac{q^2}{g}}.$$

Rearrange this for q

$$q = \left(\frac{2}{3}\right)^{3/2} g^{1/2}H^{3/2}$$

$$q = 1.71H^{3/2}.$$

This is the theoretical flow and allowance needs to be made for minor energy losses. This is usually combined with the 1.17 and introduced as a coefficient C. So

$$q = CH^{1.5}.$$

For a broad-crested weir $C = 1.6$. Here q is the discharge per unit width and so the full discharge Q is calculated by multiplying this by the length of the weir L

$$Q = CLH^{1.5}.$$

Note that strictly speaking H is the measurement from the weir crest to the energy line as it includes the kinetic energy term. In practice, H is measured from the weir crest to the water surface. The error involved in this is relatively small and can be taken into account in the value of the weir coefficient C.

As the formula is based on critical depth, it is not dependent on the shape of the weir. So the same formula can be used for any critical depth structure and not just for broad-crested weirs. Only the value of C changes to take account of the different weir shapes.

6.5.1 Determining the height of a weir

Just how high a weir must be for the flow to go critical is determined from the specific energy diagram. The effect of constructing a weir in a channel is the same as building a step up on the bed as described in Section 5.7.3. This was concerned only with looking at what happens to the depth when water flows over a step. In the case of a step-up, the depth on the step decreased and the velocity increased (Figure 6.6a). In Section 5.7.3, no thought was given to making the flow go critical. A worked example (Box 5.5) showed that for a 0.3 m high step-up, the depth of water was reduced from 0.99 m upstream to 0.67 m on the step (Figure 6.6b). This is still well above the critical depth of 0.29 m (see critical depth calculation in Box 5.7).

Now assume that the step-up on the bed is a weir and the intention is to make the flow go critical on the weir crest. This can be achieved by raising the crest level. Raising it from 0.3 m to 0.56 m (see calculation in Box 5.7) further reduces the depth on the weir from 0.67 m to 0.29 m, which is the critical depth (Figure 6.6c). This is the minimum weir height required for critical flow. Note that although the weir height has increased by 0.26 m, the upstream depth remains unchanged at 0.99 m. If the weir height is increased beyond 0.56 m, the flow will still go critical on the crest and remain at the critical depth of 0.29 m. It will not and cannot fall below this value. The difference will be in the upstream water level which will now rise. Remember there is a unique relationship between the head on a weir and the discharge. So if the weir is raised by a further 0.1 m to 0.66 m, the upstream water level will also be raised by 0.1 m to maintain the correct head on the weir (Figure 6.6d).

The operation of weirs is often misunderstood and it is believed that they cause the flow to back up and so raise water levels upstream. This only happens once critical conditions are achieved on the weir. When a weir is too low for critical flow, it is the water level on the weir which drops. The upstream level is unaffected. But once critical flow is achieved, raising the weir more than necessary will have a direct effect on the upstream water level.

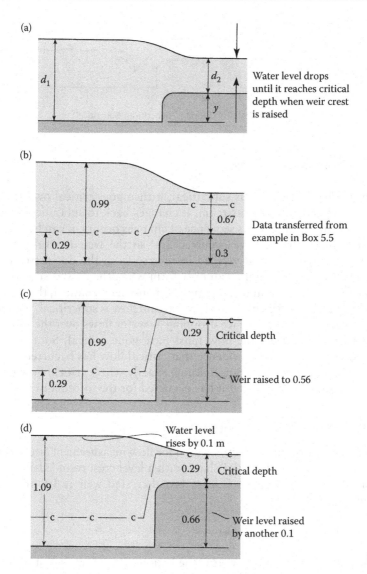

Figure 6.6 Determining the height of a weir. (a) Weir details. (b) Subcritical flow on weir. (c) Flow goes critical on weir. (d) Raising weir crest level affects upstream water level.

6.5.2 Being sure of critical flow

Critical flow must occur for the discharge formula to work. But in practice, it is not always possible to see critical flow and so some detective work is needed. Figure 6.7 shows the changing flow conditions as water flows over

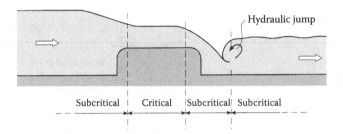

Figure 6.7 Being sure of critical flow.

a weir. Upstream the flow is subcritical, it then goes critical over the weir and then supercritical downstream. It changes back to subcritical through a hydraulic jump. When this sequence of changes occurs, it can be reasoned that critical flow must have occurred and so the weir is working properly. The changes are best verified in reverse from the downstream side. Remember a hydraulic jump can only form when the flow is supercritical, and so if there is a hydraulic jump in the downstream channel, the flow over the weir must be supercritical. If the upstream flow is subcritical, which can be verified by the water surface dropping as water flows over the weir, then somewhere in between, the flow must have gone critical. So a hydraulic jump downstream is good evidence that critical flow has occurred.

Note that it is not important to know exactly where critical flow occurs. It is enough just to know that it has occurred for the formula to work.

6.5.3 Broad-crested weirs

These are very common structures used for flow measurement (see Box 6.2). They have a broad rectangular shape with a level crest rounded at the edge (Figure 6.5b). The value of C for a broad-crested weir is 1.6 and so the formula becomes

$$Q = 1.6LH^{1.5}.$$

One disadvantage of this weir is the region of dead water just upstream. Silt and debris can accumulate here and this can seriously reduce the accuracy of the weir formula. Another is the head loss between the upstream and downstream levels. Whenever a weir (or a flume) is installed in a channel, there is always a loss of energy, particularly if there is a hydraulic jump downstream. This is the hydraulic price to be paid for measuring the flow.

6.5.4 Crump weirs

These weirs are commonly used in the United Kingdom for discharge measurement in rivers. Like the broad-crested weir, it relies on critical

---

BOX 6.2 EXAMPLE: CALCULATING DISCHARGE USING A BROAD-CRESTED WEIR

A broad-crested weir is used to measure discharge in a channel. If the weir is 2-m long and the head on the crest is 0.35 m, calculate the discharge.

The discharge over a broad-crested weir can be calculated using the formula

$$Q = 1.6LH^{1.5}.$$

Put in values for length L and head H

$$Q = 1.6 \times 2 \times 0.35^{1.5}$$

$$Q = 0.66 \text{ m}^3/\text{s}.$$

---

conditions occurring for the discharge formula to work. It has a triangular-shaped section (Figure 6.5c). The upstream slope is 1 in 2 and the downstream is 1 in 5. The sloping upstream face helps to reduce the dead water region which occurs with broad-crested weirs. It can also tolerate a high level of submergence. Its crest can also be constructed in a vee shape so that it can be used accurately for both small and large discharges.

6.5.5 Round-crested weirs

Weirs of this kind are commonly used on dam spillways (Figure 6.5a). The weir profile is carefully shaped so that it is very similar to the underside of the falling nappe of a sharp-crested weir (compare the two shapes in Figures 6.4b and 6.5a). Many standard designs are available which have been calibrated in the laboratory using hydraulic models to obtain the C values. An example is the standard weir designs produced by the US Bureau of Reclamation (see Further Reading). By constructing a weir to the dimensions given in their publications, the discharge can be measured accurately using their C values (usually between 3.0 and 4.0).

6.5.6 Drowned flow

Weirs which rely on critical depth are much less sensitive to being *drowned* (or *submerged*) than the sharp-crested type. This means that the downstream water level can rise above the weir crest without it affecting the performance of the structure, *provided* the flow still goes critical somewhere on the weir. There are ways of using weirs to measure discharge even when they are completely submerged but they are far less accurate and both upstream and downstream water depths must be measured. It is better to avoid this situation if at all possible.

6.6 FLUMES

Flumes also rely on critical flow for measuring discharges. They are sometimes called *throated flumes* because critical conditions are achieved by narrowing the width of the channel (Figure 6.8). Downstream of the throat, there is a short length of supercritical flow followed by a hydraulic jump. This returns the flow to subcritical. The formula for discharge can be determined in the same way as for solid weirs. The result is as follows

$$Q = 3.0bH^{1.5},$$

where b is width of the flume throat (m) and H is upstream depth of water (m) measured from the bed of the channel.

The head loss through flumes is much lower than for weirs and so they are ideally suited for use in channels in very flat areas where head losses need to be kept as low as possible.

6.6.1 Parshall flumes

Parshall flumes are used extensively in the United States and were developed by R. L. Parshall in 1926 (Figure 6.8b). They gained popularity in many other countries because the construction details, dimensions and calibration curves relating upstream depth to discharge have been widely published. There are several different sizes available to measure flows up to 90 m³/s. They are relatively simple to construct from a range of materials such as wood, concrete and metal because they have no curved surfaces.

If they are made and installed as recommended, they provide accurate discharge measurement.

6.6.2 WSC flumes

This is a range of standard flumes for measuring small discharges from less than 1.0 up to 50 L/s developed by Washington State College (WSC) in the United States (Figure 6.8c). They are vee-shaped so that they can be used to measure low flows accurately and, like Parshall flumes, they can be easily made up in a workshop from metal or fibre glass. They are particularly useful as portable flumes for spot measurements in small channels and irrigation furrows.

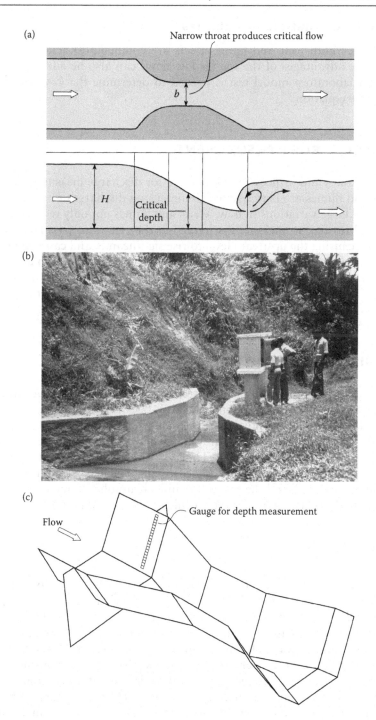

Figure 6.8 Flumes. (a) Venturi flume. (b) Parshall flume. (c) WSC flume.

6.6.3 Combination weir flumes

Sometimes both weir and flume effects are combined to achieve critical flow. The advantages of the vee for low flows can also be added. In such cases, a laboratory model test is needed to determine the C value in the discharge equation.

6.7 DISCHARGE MEASUREMENT

Weirs, flumes and orifices can all be used for discharge measurement. But weirs and flumes are better suited to measuring discharges in rivers when there are large variations in flow. Weirs and flumes not only require a simple head reading to measure discharge but they can also pass large flows without causing the upstream level to rise significantly and cause flooding. Orifice structures too can be used for flow measurement but both upstream and downstream water levels are usually required to determine discharge. Large variations in flow also mean that the gates will need constant attention for opening and closing.

6.8 DISCHARGE CONTROL

Although orifices are rather cumbersome for discharge measurement, they are very useful for discharge control. This is because the discharge through an orifice is not very sensitive to changes in upstream water level. Consider as an example, an irrigation canal system where a branch canal takes water from a main canal (Figure 6.9a). The structure at the head of the branch controls the discharge to farmers downstream. The ideal structure for this would be an orifice and the reason lies in its hydraulic characteristic curve (Figure 6.9b). This is a graph of the orifice discharge equation

$$Q = C\sqrt{2gd_1}.$$

In this simple example, the orifice opening is assumed to be fixed and so discharge Q changes only when the upstream depth d_1 changes. The point Q_d and d_d on the graph represents the normal operating condition. Now suppose the main canal operating water level rises by say 20%, because of changes in the water demand elsewhere in the system. The effect that this would have on the discharge into the branch (through the orifice) is to change it by only 5%. So even though there is a significant change in the main canal this is hardly noticed in the branch. This can be very useful for ensuring a reliable, constant flow to a farmer even though the main canal may be varying considerably due to changing demands.

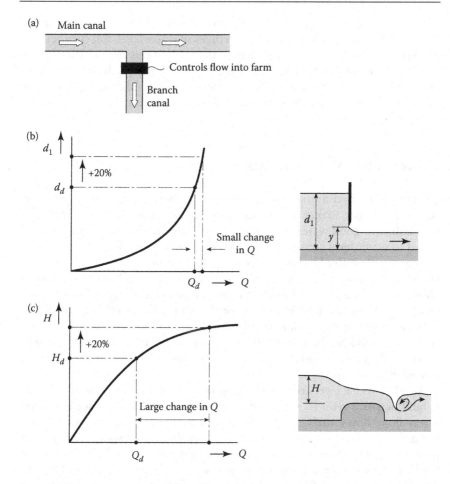

Figure 6.9 Discharge control. (a) Branch canal taking water from a main canal. (b) Using an orifice for discharge control. (c) Using a weir for discharge control.

In contrast, if a weir (or flume) is installed at the head of the branch canal, it would be very easy to use for discharge measurement but it would not be so good for controlling the flow. Again look at the hydraulic characteristic curve for a weir (Figure 6.9c). This is a graph of the weir discharge equation

$$Q = CLH^{1.5}.$$

Q_d and H_d represent the normal operating condition. Now if the water level in the main canal rises by 20%, the resulting change in discharge into the branch canal is 35%. So this type of structure is very sensitive to water level changes and would not make a very good discharge control structure.

These examples show the sensitivity of the two types of control structure. This is how they react to small changes in water level. But they refer to fixed structures. Most structures have gates (both weirs and orifices) which can be used to adjust the discharge but this adds to the burden of managing the channels. Better to use a structure which is hydraulically suited to the job to be done and which reduces the need for continual monitoring and adjustment.

6.9 WATER LEVEL CONTROL

Orifices, weirs and flumes can all be used for water level control. But the very reason that makes an orifice a good discharge regulator makes it unsuitable for good water level control. Conversely, a weir (or flume) is well suited to water level control but not for discharge control.

Imagine the water level in a river needs to be controlled to stabilise water levels in a nearby wetland site and it needs to be more or less the same both winter and summer even though the river flows change from a small flow to a flood. The structure best suited for this would be a weir as it can pass a wide range of flows with only small changes in head (water level). An orifice structure would not be suitable because large changes in water level would occur as the flow changed and the structure would require a great deal of attention and adjustment.

Another water level control problem occurs in reservoirs. When a dam is built across a river to store water, a spillway is also constructed to pass severe flood flows safely into the downstream channel to avoid overtopping the dam. The ideal structure for this is a weir because large discharges can flow over it for relatively small increases in water level. The rise in water level can be calculated using the weir discharge formula and this helps to determine the height of the dam. A freeboard is also added to the anticipated maximum water level as an added safety measure. The spillway can be the most expensive part of constructing a new dam.

Many small dams are built on small seasonal streams in developing countries to conserve water for domestic and agricultural use. But even a small dam will need a spillway, and although the stream may look small and dry up occasionally, they can suffer from very severe floods which are difficult to determine in advance, particularly when rainfall and discharge data for the stream are not available. Building a spillway for such conditions can be prohibitively expensive and far exceed the cost of the dam and so it may be cheaper in sparsely populated areas to let the dam be washed away and then rebuild it on the rare occasions that this happens. There are of course many other factors to take into account besides dam reconstruction costs such as the effects downstream of a dam break. It is a sobering thought that even the biggest and most important dams can fail because of severe flooding. They have large spillways to protect them from very severe floods which are carefully assessed at the design stage. But it is impossible to say that they will safely carry extreme floods. Nature seems to have a way of testing us by sending unexpected rain storms but fortunately these events are few and far between.

One way of avoiding the expensive spillway problem is to construct a reservoir at the side of a river rather than on the river channel itself. This is called *off-stream storage*. Water is taken from the river by gravity or by pumping into a reservoir and only a modest spillway is then needed which would have the same capacity as the inlet discharge. When a flood comes down the river, there is no obstruction in its path and so it flows safely passing the reservoir.

6.10 ENERGY DISSIPATORS

6.10.1 Stilling basins

When water flows over a weir or through a flume and becomes supercritical, it can do a lot of damage to the downstream channel if it is left unprotected. This is particularly true when water rushes down an overflow spillway on a dam. The water can reach very high velocities by the time it gets to the bottom. Scour can be prevented by lining the channel but this can be an expensive option.

The alternative is to convert the flow to subcritical using a hydraulic jump. The requirements for a jump are supercritical flow upstream and subcritical flow in the downstream channel. The main problem is to create the right flow conditions in the downstream channel for a jump to occur even though the discharge may range from a small overspill flow to a large flood. Consider what happens when flow reaches the bottom of a spillway. If the tailwater is too shallow for a jump to form, the supercritical flow will shoot off downstream and no jump will be formed. If the water is too deep, the jump will be drowned and the supercritical flow will rush underneath and still cause erosion for some distance downstream. These problems can be resolved by building *stilling basins* which create and confine hydraulic jumps even though the tailwater may not be at the ideal depth for a jump to occur naturally. There are many different designs available but perhaps the simplest is a small vertical wall placed across the channel (Figure 6.10a). Other more sophisticated designs have been developed in laboratories using models as it is not possible to design them using formulae. The choice of stilling basin is linked to the Froude number of the upstream supercritical flow (Figure 6.10b).

6.10.2 Drop structures

Drop structures are used to take flow down steep slopes step by step to dissipate energy and so avoid erosion (Figure 6.11). Channels on steep sloping land are prone to erosion because of the high velocities. One option to avoid

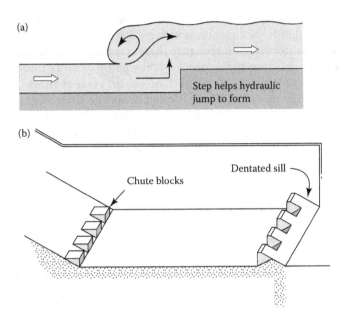

Figure 6.10 Stilling basins. (a) Step to form hydraulic jump. (b) Stilling basin details.

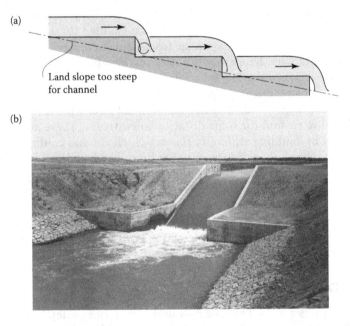

Figure 6.11 Drop structures. (a) Drop structures slow flow down steep slope. (b) Drop structure on an irrigation canal in Iraq.

this is to line channels with concrete or brick but another is to construct natural channels on a gentle gradient and to build in steps like a staircase. The water flows gently and safely along the shallow reaches of channel and then drops to the next reach through a drop structure. Often a drop is made into a weir so that it can also be used for discharge measurement. It can also be fitted with gates so that it can be used for water level and discharge control. Drop structures are usually combined with stilling basins so that unwanted energy is got rid of effectively.

6.11 SIPHONS

Siphons are hydraulic structures which have always fascinated engineers. Used long ago by ancient Greeks and Egyptians, they began to be used seriously in civil engineering in the mid-nineteenth century as spillways on storage reservoirs. More recently, their special characteristics have been put to use in providing protection against sudden surges in hydropower intakes and in controlling water levels in rivers and canals subjected to flooding.

Although siphons have been successfully installed in many parts of the world, there is still a general lack of guidelines for designers. This is borne out by the wide variety of siphon shapes and sizes that have been used. Invariably, design is based on intuition and experience of previous siphon structures, and few engineers would attempt to install such a structure without first carrying out a model study.

The principle of siphon operation is described in Section 4.7 in connection with pipe flow. In its very simplest form, it is a pipe which rises above the hydraulic gradient over part of its length. The following siphon structures, although more sophisticated, still follow this same principle.

6.11.1 Black-water siphons

These are the most common type of siphon, and Figure 6.12 shows how one can be used as a spillway from a reservoir. It consists of an enclosed barrel which is sealed from the atmosphere by the upstream and downstream water levels. The lower part is shaped like a weir and the upper part forms the hood.

Water starts to flow through a siphon when the upstream water level rises above the crest. As the flow plunges into the downstream water, it entrains and removes air from inside the barrel. As the barrel is sealed, air cannot enter from outside and so the pressure gradually falls and this increases the flow rate until the barrel is running full of water. At this stage, the siphon is said to be *primed* and flow is described as *black-water flow*. This is in contrast to the flow just before priming when there is a lot of air entrained and it has a white appearance. This is termed *white-water flow*.

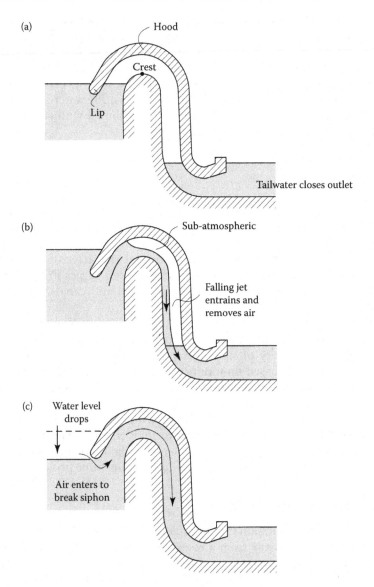

(a)

Hood

Crest

Lip

Tailwater closes outlet

(b)

Sub-atmospheric

Falling jet
entrains and
removes air

(c)   Water level
drops

Air enters to
break siphon

Figure 6.12 Black-water siphon. (a) Typical siphon design. (b) Priming. (c) Depriming.

Priming takes place rapidly until the discharge reaches the siphon's full capacity. The discharge is determined by the size of the barrel and the difference in energy available across the siphon. The siphon eventually starts to draw down the upstream water level and this continues even when the water level falls below the crest. Only when it is drawn down to the lip on the hood on the inlet, can air enter the barrel and break the siphonic action. Flow then stops rapidly and will only begin again when the siphon is reprimed.

Black-water siphons have several advantages over conventional spillways. There are no moving parts such as gates and so they do not need constant attention from operators. They respond automatically, and rapidly, to changes in flow and water levels and so floods which come unexpectedly in the night do not cause problems. Their compactness also means that they are very useful when crest lengths for conventional weir spillways are limited. But they are not without problems. The abruptness of priming produces a sudden rush of water which can cause problems downstream. They are also prone to *hunting* when the flow towards a siphon is less than the siphon capacity. They are continually priming and de-priming and this can cause surges downstream and vibration which is not good for the structure.

6.11.2 Air-regulated siphons

This type of siphon is a more recent development and offers many advantages over the more common black-water siphon. It automatically adjusts its discharge to match the approach flow and at the same time maintains a constant water level on the upstream side. This is achieved by the siphon passing a mixture of air and water.

Air-regulated siphons are typically used for water level control in reservoirs and in rivers and canals. They will maintain a constant water level in a channel even though the discharge is changing. This would be an ideal structure for the wetland example discussed earlier.

Figure 6.13a shows a typical air-regulated siphon for a river. There needs to be sufficient energy available to entrain air in the barrel and take it out to prime the siphon. This one is designed to operate at very low heads of 1–2 m (difference between the upstream and downstream water levels). The siphon is shaped in many ways like a black-water siphon and relies on the barrel being enclosed and sealed by the upstream and downstream water levels. But the main difference is the inlet to the hood or upstream lip. This is set above the crest level, whereas in a black-water siphon it is set below. A step is also included in the down-leg to encourage turbulence and air entrainment.

The operation of the siphon has several distinct phases (Figure 6.13a–e):

Phase I: Free weir flow (Figure 6.13a). As the upstream water level starts to rise due to increased flow, water flows over the crest and plunges into the downstream pool. The structure behaves as a conventional free-flowing weir. As the water level has not yet reached the upstream lip, air which is evacuated by the flow is immediately replaced and the pressure in the barrel remains atmospheric.

Phase II: Deflected flow (Figure 6.13b,c). As the flow increases, the upstream water level rises further and seals the barrel. The evacuation

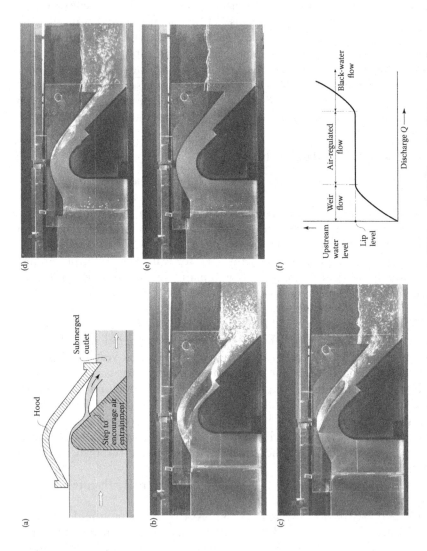

Figure 6.13 Air-regulated siphon. (a) Phase I. (b) Phase II begins. (c) Phase II. (d) Phase III. (e) Phase IV. (f) Discharge characteristic.

of air continues and a partial vacuum is created. This raises the head of water over the crest and so increases the discharge through the siphon. A point is reached when the flow through the siphon exceeds the incoming flow. The water surface close to the lip is drawn down and air is sucked into the barrel to compensate for the evacuation taking place on the downstream side. This process is not cyclic but continuous. Both air and water are drawn continuously through the structure. In this manner, the siphon adjusts rapidly and smoothly to the incoming flow and is said to be self-regulating. As the flow passes over the weir, it is deflected by a step and springs clear of the structure. This encourages air entrainment and evacuation at low flows and it not intended to create an air seal as in the case of some black-water siphons.

Phase III: Air-partialised flow (Figure 6.13d). As the flow increases, the water level inside the hood rises to a point where the air pocket is completely swept out and the siphon barrel is occupied by a mixture of air and water. Changes in the discharge are now accommodated by variations in the quantity of air passing through the siphon and not by an increase in the effective head over the weir crest.

Phase IV: Black-water flow (Figure 6.13e). Increasing the flow beyond the air-partialised phase produces the more common black-water flow in which the barrel is completely filled with water. The discharge is now determined by the head across the siphon, that is, the difference between the upstream and downstream water levels.

The flow changes from one phase to another quite gradually and smoothly and there is no distinct or abrupt change-over point. During phases II and III, the upstream water level remains relatively constant at a level close to that of the upstream lip. This feature makes this structure an excellent water level regulator. Only when the black-water phase is reached, does an increase in discharge cause a significant rise in the upstream water level (Figure 6.13e).

6.12 CULVERTS

Culverts are very useful structures for taking water under roads and railways. They are circular or rectangular in shape and their size is chosen so that they are large enough to carry a given discharge, usually with minimum energy loss (Figure 6.14). They are important structures and can be as much as 15% of the cost of building a new road.

Although they are very simple in appearance, culverts can be quite complex hydraulically depending on how and where they are used. Sometimes they flow part full, like an open channel and other times they can flow full, like a pipe. Six different flow conditions are recognised depending on

Figure 6.14 Culverts.

the size and shape of a culvert, its length and its position in relation to the upstream and downstream water level (Figure 6.15).

The simplest condition occurs when a culvert is set well below both the upstream and downstream water levels and it runs full of water (type I). It behaves like a pipe and the difference between the water levels is the energy available which determines the discharge. This full pipe flow can still occur even when the downstream water level falls below the culvert soffit (this is the roof of the culvert, type II). The culvert is now behaving like an orifice and the flow is controlled by the upstream water level and the size of the culvert. In both cases, the discharge is controlled by what is happening downstream. When the water level falls, the discharge increases and when it rises, the discharge decreases. If the flow in the channel does not change, then any change in the downstream water level will have a corresponding effect on the upstream water level.

The four remaining flow conditions are for open-channel flow. Three occur when both the upstream and downstream water levels are below the culvert soffit (type III, IV and V). The difference between III and IV is the slope of the culvert. Type III produces subcritical flow and so the discharge is controlled by the downstream water level. When the downstream level rises, the discharge reduces or the upstream level rises to accommodate the same flow. In IV, the flow is still subcritical but the downstream water level is low and so the flow goes through the critical depth at the outlet. This means that the flow is controlled by the upstream level. Because the flow has gone through the critical depth, any changes downstream do not affect the flow in the culvert or upstream. Condition V produces super-critical flow and so the culvert is again controlled by the upstream level. A rise in the downstream level will have no effect until it starts to drown the culvert.

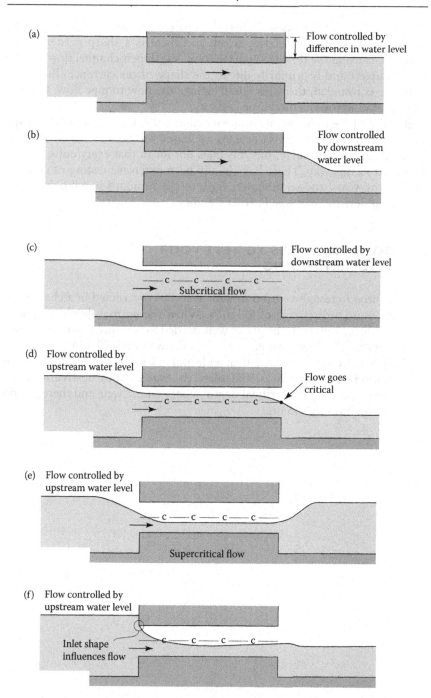

Figure 6.15 Culvert flow conditions. (a) Type I. (b) Type II. (c) Type III. (d) Type IV. (e) Type V. (f) Type VI.

The final condition occurs when the upstream water level is above the soffit, the downstream level is below and there is a sharp corner at the entrance causing the flow to separate (type VI). Open-channel flow occurs in the culvert and is primarily due to the shape of the entrance. But if the entrance is rounded, then this could change the flow to pipe flow. Clearly, this would improve its discharge capacity.

Because of their complexity and particularly the importance of the shape of the entrance, most culvert designs are based on model tests rather than on fundamental formulae. But this does not mean that every culvert must be tested in this way. There has already been extensive testing of culverts over many years and so designers look to the standard handbooks when designing new culverts (see Further Reading).

6.13 SOME EXAMPLES TO TEST YOUR UNDERSTANDING

1. A broad-crested weir 3.5 m wide is to be constructed in a channel to measure a discharge of 4.3 m³/s. When the normal depth of flow is 1.2 m, calculate the height of weir needed to measure this flow assuming that the flow must go critical on the weir crest (0.4 m).

2. A broad-crested weir is 5.0-m wide and 0.5-m high is used to measure a discharge of 7.5 m³/s. Calculate the water depth upstream of the weir assuming that critical depth occurs on the weir and there are no energy losses (1.45 m).

Chapter 7

Pumps

7.1 INTRODUCTION

Few water supply systems have the benefit of a gravity supply, most require some kind of pumping. Pumps are a means of adding energy to water. They convert fuel energy, such as petrol or diesel, into useful water energy using combustion engines or electric motors. In the typical pipe flow problem in Chapter 5, the energy to drive water along the pipeline to the town was from a reservoir located high above the town. The energy line drawn from the reservoir to the town indicated the amount of energy available. Adding a pump to this system increases the available energy and raises the energy line so that the discharge from the reservoir to the town can be increased (Figure 7.1).

Pumps have been used for thousands of years. Early examples were largely small hand or animal-powered pumps for lifting small quantities of water. It was not until the advent of the steam engine, only two centuries ago, that the larger rotating pumps were developed and became an important part of the study of hydraulics. Consequently, there are two main types of pump: *positive displacement pumps*, which are mostly small hand and animal-powered pumps still in common use in developing countries; and *roto-dynamic pumps*, those driven by diesel or electric motors and used in all modern water supply and irrigation systems.

Because of the importance of pumping, most of this chapter is devoted to pumps, but mention is also made of turbines which are hydraulic machines for generating energy.

7.2 POSITIVE DISPLACEMENT PUMPS

Positive displacement pumps usually deliver small discharges over a wide range of pumping heads. Typical examples are hand-piston pumps, rotary pumps, airlift pumps and Archimedean screws (Figure 7.2).

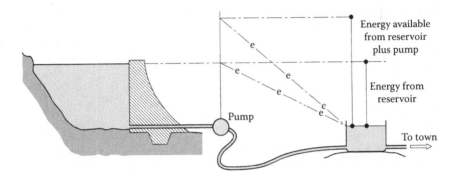

Figure 7.1 Pumps add energy to pipe systems.

7.2.1 Typical pumps

Hand-piston pumps (Figure 7.2a and e) are used extensively in developing countries for lifting groundwater for domestic water supplies. A pipe connects the pump to the water source – usually a well. At the end of this pipe is a non-return valve that only allows water to enter the pump and stops it from flowing back into the source. The pump itself comprises a piston and cylinder. These must have a very close fit so that when the piston is raised, it creates a vacuum in the cylinder and water is drawn up into the pump. When the piston is pushed down, the water is pushed through a small valve in the piston to fill up the space above it. As the piston is raised again, it lifts the water above it so it pours out through the spout of the pump and into a tank or other water-collecting device. The procedure is then repeated. The discharge from the pump depends on the energy available from those working the pump handle. You basically get out what you put in. The height to which water can be pumped in this way is fixed only by the strength of those pumping and the pump seals, which will start to leak when the pressure gets too high.

The rotary pump (Figure 7.2b) contains two gears that mesh together as they rotate in opposite directions. The liquid becomes trapped between the gears and is forced into the delivery pipe.

The airlift pump (Figure 7.2c) uses an air compressor to force air down a pipe into the inlet of the water pipe. The mixture of air and water, which is less dense than the surrounding water in the well, then rises up above ground level. This pump is not very efficient but it will pump water that has sand and grit in it which would normally damage other pumps.

The Archimedean screw pump (Figure 7.2d) has been used for thousands of years and is still used today for pumping irrigation water in Egypt. It comprises a helical screw inside a casing. Water is lifted by turning the screw by hand. Some modern pumping stations also use this idea but the screws are much larger and are driven by a diesel or electric power unit.

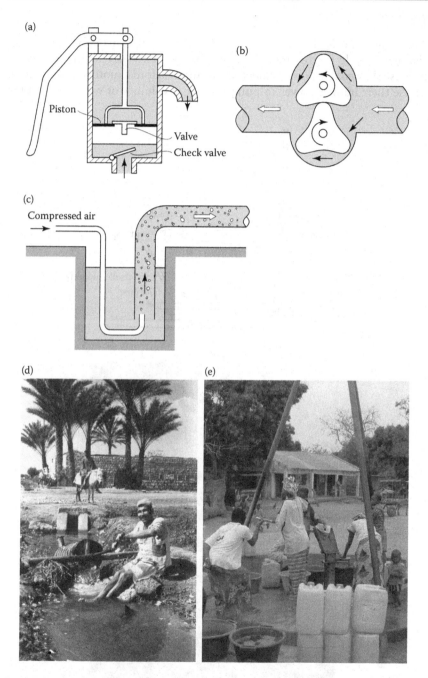

Figure 7.2 Positive displacement pumps. (a) Piston pump. (b) Rotary pump. (c) Airlift
pump. (d) Archimedean screw pump in Egypt. (e) Hand pump in village in the
Gambia.

7.2.2 Treadle pumps

The treadle pump is another example of a positive displacement pump (Figure 7.3). It is growing in popularity in many developing countries because they enable poor farmers, who cannot afford a motor-driven pump, to use their human power to pump the large volumes of water needed for

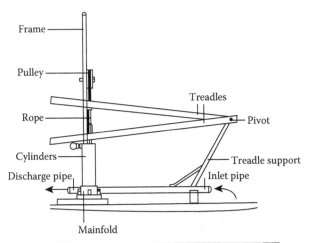

Figure 7.3 Treadle pump.

irrigating crops. It was first developed in Bangladesh in the early 1980s for irrigating flooded rice and it is now estimated that some 2 million pumps are in use throughout Asia. Their popularity has spread in recent years to sub-Saharan Africa. The designers of the treadle pump have skillfully adapted the principle of the hand-piston pump (described above) so that greater volumes of water can be lifted. Two pistons and cylinders are used to raise water instead of one. But the most important innovation was the change in driving power from arms and hands to legs and feet. Leg muscles are much more powerful than arm muscles and so they are capable of lifting much more water. The two cylinders are located side by side and a chain or rope, which passes over a pulley, connects the two pistons together so that when one piston is being pushed down, the other one is being raised. Each piston is connected to a treadle on which the operator stands and pushes them up and down in a rhythmic motion – like pushing peddles on a bicycle. This rhythmic method of driving the pump seems to have gained wide acceptance among farmers in the developing world.

Although the pump is popular in several developing countries, some exaggerated claims are often made about its physical performance usually by those who have little knowledge of how the physical world works. Some claim it will reach pressures up to 15 m and deliver discharges of 1–2 L/s and more as if it contains magical properties. Applying some simple hydraulic principles to this pump can dispel some of this magic and show just what it is capable of producing for a small farmer in a developing country (see Box 7.3). The reality is more like 0.6 to 1.2 L/s with a head of 1–2 m. Like most things in life, you can only get out of this pump what you are prepared to put into it in terms of power and energy – I am afraid there is no 'free lunch' even in pumping.

Typically, one fit, male operator could irrigate up to 0.25 ha which can dramatically transform a family's livelihood from subsistence to one where there is a cash income from irrigated produce.

7.2.3 The heart: an important pump

One final positive displacement pump worth mentioning is the heart which pushes blood around your body. The muscles around the heart contract and push blood into your arteries, which is your body's pipe system. It must create enough pressure to force blood to flow around your entire body with enough flow to carry all the nutrients your body needs. Problems of high blood pressure (known as hypertension) occur when the arteries (the pipe system) become restricted in some way. Your heart then increases the pressure to maintain circulation. Applying basic hydraulics to this, fatty deposits in your pipes (or arteries) increase pipe friction and can also reduce pipe diameters. Muscles surrounding the pipes are also known to contract and reduce pipe diameters. All these physical factors can push up the pumping pressure. Also the taller you are, the more pressure your heart needs to produce to get blood to the top of your head and lift it back up from your feet to your heart. Flow in arteries is mostly laminar, although in the larger main artery near the heart it can be turbulent. This implies that friction losses in arteries is a function of viscosity rather than roughness and so anything that 'thins' blood (reduces it viscosity) should ease the flow. Some drugs are known to do this as well as others that stop blood clots and pipe blockages. Blood pressure-reducing drugs work in many different ways. Some slow down the heart rate and others increase pipe diameters by reducing the squeezing effects of the muscles around them. But the process is not simply a physical one. It is complex mixture of physical and biological processes and there is still a great deal not yet known about how these processes combine and work together. But basic hydraulics undoubtedly plays a part.

7.3 ROTO-DYNAMIC PUMPS

All roto-dynamic pumps rely on spinning an impeller or rotor for their pumping action. There are three main types that are described by the way in which water flows through them:

- Centrifugal pumps
- Mixed flow pumps
- Axial flow pumps

7.3.1 Centrifugal pumps

Centrifugal pumps are the most widely used of all the roto-dynamic pumps. The discharges and pressures they produce are ideally suited to water supply and irrigation schemes.

To understand how centrifugal pumps work, consider first how centrifugal forces occur. Most people will, at some time, have spun a bucket of water around at arm's length and observed that water stays in the bucket even when it is upside down (Figure 7.4a). Water is held in the bucket by the centrifugal force created by spinning the bucket. The faster the bucket is spun, the tighter the water is held. Centrifugal pumps make use of this idea (Figure 7.4b). The bucket is replaced by an *impeller* which spins at high speed inside a spiral casing. Water is drawn into the pump from the source of supply through a short length of pipe called the *suction*. As the impeller spins, water is thrown outwards and is collected by a spiral-shaped pump casing and guided towards the outlet. This is the *delivery* side of the pump.

The design of the impeller and the casing is important for efficient pump performance. As water enters the pump through the suction pipe, the opening is small and so velocity is high, and as a consequence of this, the pressure is low (think about the energy equation). As the flow moves through

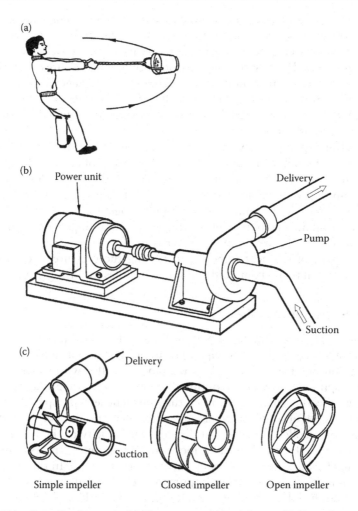

Figure 7.4   Centrifugal pumps. (a) Creating centrifugal force. (b) Typical centrifugal pump-set. (c) Different types of pump impellers.

the pump and up between the blades of the impeller, the flow gradually expands. This causes the velocity to fall and the pressure to rise again. It is important at this stage to recover as much pressure energy as possible and not to lose energy through friction and turbulence. This is achieved by carefully shaping the impeller blades and the spiral pump casing so that the movement of water is as 'streamline' as possible. Any sudden change in change in shape would create a great deal of turbulence, which would dissipate the water energy instead of increasing the pressure.

Some pumps have very simple impellers with straight blades (Figure 7.4c), but these create a lot of turbulence in the flow and so pressure recovery is not so good. Energy losses tend to be high and the efficiency is poor. From a practical point of view, pumps like this are cheap to make and are used where efficiency is not so important such as in domestic washing machines. Larger pumps need more careful design and have curved vanes so that the water enters and leaves the impeller smoothly. This ensures that energy losses are kept to a minimum and a high level of efficiency of energy and power use can be achieved. Most impellers have side plates and are called *closed impellers*. When there is debris in the water, *open impellers* are used to reduce the risk of blockage.

Centrifugal pumps are very versatile and can be used for a wide variety of applications. They can deliver water at low heads of just a few metres up to 100 m or more. The discharge range is also high, from a few litres per second up to several cubic metres per second. Higher discharges and pressures are achieved by running several pumps together (see Section 7.10) or by using a multistage pump. This comprises several impellers on the same shaft, driven by the same motor. Water is fed from the outlet of one stage into the inlet of the next impeller, increasing the pressure at each stage. Multistage pumps are commonly used in boreholes for lifting groundwater.

7.3.2 Some pump history

One of the earliest centrifugal pumps was developed in the late eighteenth century and used a spinning pipe that discharged into a semicircular collecting channel (Figure 7.5a). Once the pipe was filled with water (primed) and was spinning, water was continually drawn up the tube and discharged into the collecting channel. Another, the Massachusetts pump built in 1818, comprised a set of simple straight blades mounted on a horizontal shaft surrounded by a collecting case that directed the velocity of the water up the discharge pipe (Figure 7.5b). This looks similar in some respects to modern centrifugal pumps but it differs in one important aspect. In this and other earlier pumps, the water leaves the pump impeller at high speed and friction is left to do the job of slowing down the water once it leaves the pump. This

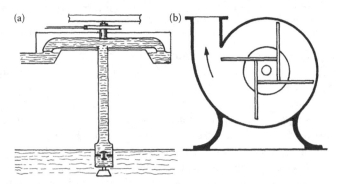

Figure 7.5 Early examples of roto-dynamic pumps. (a) Water discharges from pipe into collecting channel. (b) Simple straight bladed impeller.

means that they were not very efficient at transferring energy and power to the water and so their performance was rather poor.

It was the application of the energy equation to pumps that revolutionised pump design and performance. This was the realisation that when water is flowing in a pipe, pressure can be converted to velocity and vice versa. It is well known that pressure is needed to produce a high-velocity jet of water but what was not so obvious was that the pressure can be recovered if the water jet is slowed down in a smooth manner. This was the understanding that was missing in the early pumps. Designers knew that the spinning action would speed up the water but they did not realise that pressure could be recovered again by smoothly slowing the water down. It is the recovery of pressure energy that marks out the design of modern pumps.

7.3.3 Axial flow pumps

Axial flow pumps consist of a propeller housed inside a tube that acts as a discharge pipe (Figure 7.6). The power unit turns the propeller by means of a long shaft running down the middle of the pipe and this lifts the water up the pipe. These pumps are very efficient at lifting large volumes of water at low pressure. They are ideally suited for lifting water from rivers or lakes into canals for irrigation and for land drainage where large volumes of

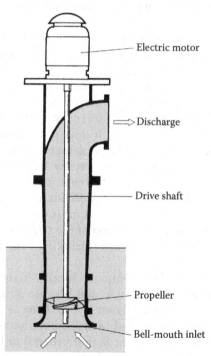

Figure 7.6 Axial flow pumps.

water need to be lifted through a few metres. However, they tend to be very expensive because of the high cost of materials, particularly the drive shaft and bearings to support the shafted propeller. For this reason, they tend only to be used for large pumping works.

7.3.4 Mixed flow pumps

These pumps are a mixture of axial flow and centrifugal pumps and so combine the best features of both pump types. Mixed flow pumps are more efficient at pumping larger quantities of water than centrifugal pumps and are more efficient at pumping to higher pressures than axial flow pumps.

7.4 PUMPING PRESSURE

7.4.1 Suction lift

Pumps are usually located above the water source for the convenience of the users and this leads to the idea that pumps must 'suck' up water from the source before lifting it to some higher level. But this idea and the word 'sucking' can lead to misunderstandings about how pumps work (Figure 7.7).

'Sucking' water up a drinking straw is very similar to the way a pump draws water from a source. However, you do not actually suck up the water, you suck out the air from the straw and create a vacuum. Atmospheric pressure does the rest. It pushes down on the water surface in the glass and forces water up the straw to fill the vacuum. So atmospheric pressure provides the driving force but it also puts a limit on how high water can be lifted in this way. It does not depend on the ability of the sucker. At sea level, atmospheric pressure is approximately 10 m head of water and so if you were relying on a straw 10.1 m long for your water needs, you would surely die of thirst! Even a straw 9 m long would cause you a lot of problems. It would be very difficult to maintain the vacuum in the straw and you would probably spend most of the day taking the small trickle of water that emerged at the top of the straw just to stay alive. The shorter the straw, however, the easier drinking becomes, and the more water you get.

This same principle applies to pumps. Ideally, it should be possible to draw water from a source up to 10 m. But because of friction losses in the pipework, an upper limit is usually set at 7 m. Even at this level, there can be problems in keeping out air and maintaining a vacuum and so a more practical limit is 3 m. When water is more than 3 m below the pump, a shelf can be excavated below ground level to bring the pump closer to the water. For pumps operating at high altitudes in mountainous regions where

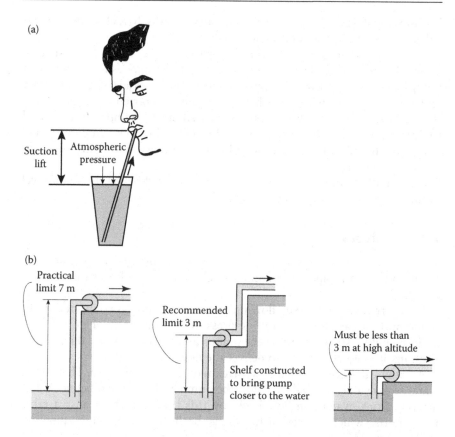

Figure 7.7 Pump suction. (a) Creating a vacuum in a straw. (b) Practical limits for suction lift.

atmospheric pressure is less than 10 m, the practical limit of 3 m will need to be lowered further for satisfactory pump operation.

The difference in height between the water source (usually referred to as the sump) and the pump is called the *suction lift*. It is an unfortunate name as it clearly does not describe what actually happens in practice. However, it is in common use and so we are stuck with.

Not all pumps suffer from suction problems. Some pumps are designed to work below the water level in the sump and are called *submersible pumps*. These are often used for deep boreholes and are driven by an electric motor, which is also submerged, connected directly to the pump drive shaft. The motor is well sealed from the water but being submerged helps to keep the motor from overheating.

Excessive suction lift can affect pump performance and this is demonstrated by the results of a test performed on a pump for a small rural water supply scheme. The pump delivered 6.5 L/s with a 3-m suction lift. But when the suction lift was increased to 8 m, the discharge dropped to 1.2 L/s – a loss in flow of 5.3 L/s – only 15% of the original discharge! So the general rule is to keep the suction lift as short as possible.

Centrifugal pumps will not normally suck out the air from the suction and will only work when the pump and the suction pipe is full of water. If pumps are located above the sump, then it will be necessary to fill the pump and the suction pipe with water before the pump is started up. This process is called *priming*. A small handpump, located on the pump casing, is used to evacuate the air. When pumps are located below the sump, they will naturally fill with water under the influence of gravity and so no additional priming is required.

For axial flow pumps, the impeller is best located below the water level in the sump and so no priming is needed.

7.4.2 Delivery

The delivery side of a pump comprises pipes and fittings that connect the pump to the main pipe system and provide some control over the pressure and discharge.

For centrifugal and mixed flow pumps, a sluice valve is connected to the pump outlet to assist in controlling pressure and discharge (Figure 7.8). It is closed before starting so that the pump can be primed. Once the pump is running, it is slowly opened to deliver the flow.

A reflux valve is connected downstream of the delivery valve. This allows water to flow one way only – out of the pump and into the pipeline. When a pump stops, water can flow back towards the pump and cause a rapid pressure rise which can seriously damage both the pump and the pipeline. (see Section 7.14). The reflux valve prevents the return flow from reaching

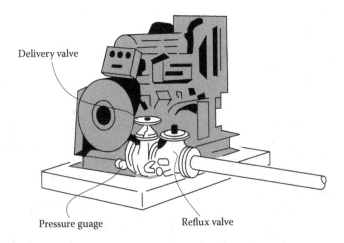

Delivery valve

Pressure guage   Reflux valve

Figure 7.8 Delivery side of a centrifugal pump.

the pump. Some reflux valves have a small by-pass valve fitted which allows water stored in the pipe to pass around the valve and to prime the pump.

7.4.3 Pumping head

Pumping head is another term that is widely misunderstood. It is often confused with the *delivery lift (or head)* – the pressure created on the delivery side of the pump. Pumping water requires energy not only to deliver water but also to draw it up from its source. So the pumping head is the sum of the delivery lift and suction lift and the friction losses in both suction and delivery pipes (Figure 7.9).

pumping head (m) = suction lift (m)

+ friction loss in suction (m)

+ delivery lift (m)

+ friction loss in pipeline (m).

The suction lift refers to the elevation change between the sump water level and the pump, and the delivery lift is the elevation change between the pump and the point of supply, in this case, it is the reservoir water level. Both are fixed values for a given installation. The friction losses in the sump and the delivery will vary depending on the discharge. There are also minor losses at the inlet and outlet to a pipeline but these tend to be very small in comparison to the main lift and other friction loses. To allow for all the minor losses, it is common practice to add 10% to the pumping head rather than to try and work them out in detail.

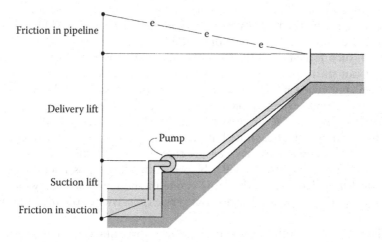

Figure 7.9 Pumping head.

So if the suction lift is 4 m and the pump then delivers 7 m head, the pumping head would be 11 m. This represents the total height through which the water must be lifted from source to delivery point. If 11 m was the maximum that a pump can deliver, then any change in the suction lift would affect the delivery lift. For example, if the suction lift increased to 6 m, then the delivery lift would reduce to 5 m resulting in the same overall total pumping head of 11 m. So just quoting delivery lift without any reference to suction lift does not provide enough information about what a pump can do in pressure terms.

In many pumping installations, the pumping head also includes any losses in head resulting from friction in the suction pipes and the losses as the water flows through filters and valves, and also friction and fittings losses on the delivery side.

7.4.4 Cavitation

A particular problem associated with excessive suction lift is a phenomenon known as *cavitation* which can not only damage the pump but also reduce its operating efficiency (see also Section 3.9.2). When water enters a pump near the centre of the impeller, the water velocity is high and so the pressure is very low. This is exacerbated when the suction lift is high. In some cases, the pressure is so low that it reaches the vapour pressure of water (approximately 0.5 m absolute) and cavities (similar to bubbles) begin to form in the flow. The cavities are very small – less than 0.5-mm diameter – but there are usually so many of them that the water looks milky in appearance. This must not be confused with air entrainment which looks similar. This occurs when there is a lot of turbulence just downstream of a weir or a waterfall and air and water are mixed to produce a milky looking fluid. Cavities are not air-filled bubbles. Rather they are 'holes' in the water that contain only water vapour. The problems occur when the cavities move through the pump into an area of higher pressure where they become unstable and begin to collapse (Figure 7.10a). A small needle jet of water rushes across the cavity with such force that if the cavity is close to the pump impeller or casing then it can start to damage them. Each cavity collapse is not significant on its own, but when many thousands of them continually collapse close to the metal surfaces, then damage begins to grow. In some cases, it has been known to completely wear away the impeller blades (Figure 7.10b).

Most pumps in fact cavitate but not all cavitation causes damage. Pump designers go to great lengths to ensure that the cavities collapse in the main flow, well away from the impeller blades. If this is not possible, then stainless steel impellers are used as this is one of the few materials which can resist cavitation damage. But this is a very expensive option and not one to take up lightly.

Cavitation not only erodes the surfaces in contact with water but it can also be very noisy. It also causes vibration and can reduce pumping

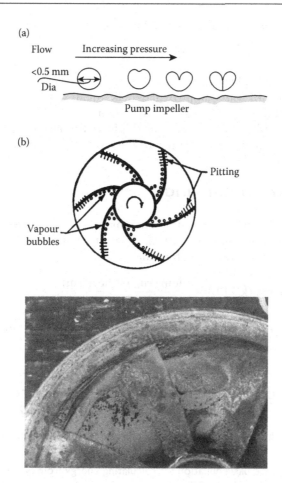

(a)

Flow   Increasing pressure

<0.5 mm
Dia

Pump impeller

(b)

Pitting

Vapour
bubbles

Figure 7.10 Cavitation. (a) How cavitation bubbles collapse. (b) Cavitation damage to the blades of a pump impeller.

efficiency. If you have an opportunity to visit a water pumping station, then place your ear against the pump casing and you will hear the cavitation. Above all the noise of the pump engines, it should be possible to hear a sharp crackling sound. This is the sound of the cavities collapsing as the pressure rises in the pump. The level of the noise produced by the cavitation is a measure of how badly the pump is cavitating.

Cavitation also occurs in turbine runners which are discussed briefly later in this chapter. High velocities at the turbine inlet produce cavities which then collapse close to the runner blades near the exit.

A concept sometimes used for limiting suction lift and so avoiding cavitation is the *net positive suction head* (*NPSH*). Manufacturers normally specify a value of NPSH for each discharge to ensure a pump operates satisfactorily.

The NPSH for an installation can be calculated as follows:

$$\text{NPSH} = P_a - H_s - V_p - h_f,$$

where P_a is atmospheric pressure (m), H_s is suction head (m), V_p is the vapour pressure of water (approximately 0.5 m absolute) and h_f is the head loss in the suction pipe (approximately 0.75 m).

The value of NPSH calculated using this formula must be greater than that quoted by the pump manufacturer for the pump to operate satisfactorily.

7.5 ENERGY FOR PUMPING

Energy is needed to pump water. The amount required depends on both the volume of water pumped and the height to which it is lifted. It can be calculated using the following formula:

$$\text{water energy (kWh)} = \frac{\text{volume (m}^3) \times \text{head (m)}}{367}.$$

A derivation of this formula is shown in Box 7.1 for the more inquisitive reader.

BOX 7.1   DERIVATION: A FORMULA TO CALCULATE ENERGY FOR PUMPING

Derive a formula to calculate the amount of energy needed to lift and fill an elevated tank with water (Figure 7.12).

Start with the basic equation for calculating energy from Section 1.9:

water energy = work done.

So

water energy (Nm) = force (N) × distance (m).

We now need to establish what the values of force and distance mean in terms of pumping water.

Look first at force. This is the total weight of water to be lifted. Usually, the volume is known and so we use Newton's second law to determine the weight:

force = mass (kg) × acceleration (m/s^2).

In this case, acceleration is that due to gravity and so:

mass of water = volume (m^3) × density (kg/m^3).

And:

force = volume (m^3) × density (kg/m^3) × gravity constant (m/s^2).

Now density ρ = 1,000 kg/m^3 and the gravity constant g = 9.81 m/s^2 so:

force = volume (m^3) × 1,000 × 9.81
      = volume (m^3) × 9,810.

The 'distance' part of the energy equation is the pumping head and so:

distance = pumping head (m).

Putting the values for force and distance in the energy equation:

water energy (Nm) = volume (m^3) × 9,810 × head (m).

But this is in Nm. We need to convert this to more useful and practical units of energy like kWh

1 kWh = 3,600,000 Nm (section 1.9.2).

And so

$$\text{water energy (kWh)} = \frac{\text{volume (m}^3\text{)} \times 9,810 \times \text{head (m)}}{3,600,000}.$$

This simplifies to the formula which is most often used:

$$\text{water energy (kWh)} = \frac{\text{volume (m}^3\text{)} \times \text{head (m)}}{367}.$$

An example of how to use this formula for calculating energy for pumping is shown in Box 7.2.

7.6 POWER FOR PUMPING

The terms power and energy are often used to mean the same thing. But they are different. Power is the rate of energy use. So energy is determined by volume and head, whereas power is determined by discharge and head. Power is usually measured in Watts (W) but this is a relatively small amount of power and so it is more common to speak of pumping power in terms of kilowatts (kW).

One way of calculating power is to divide energy by the time taken to use the energy. So

$$\text{water power (kW)} = \frac{\text{energy (kWh)}}{\text{time (h)}}.$$

But another way of calculating power is to use the formula

$$\text{water power (kW)} = 9.81 \times \text{discharge (m}^3\text{/s)} \times \text{pumping head (m)}$$
$$= 9.81 \, QH.$$

The derivation of this formula is similar to that of water energy but discharge is used instead of volume.

A graph of water power and discharge for a wide range of pumping heads is a convenient way of showing the range of this formula (Figure 7.11).

Manufacturers often quote discharges in m³/h rather than in m³/s. In this case, the formula becomes

$$\text{water power (kW)} = \frac{\text{discharge (m}^3\text{/h)} \times \text{head (m)}}{367}.$$

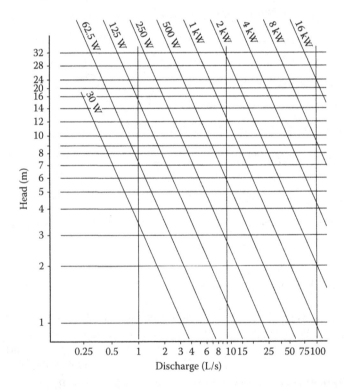

Figure 7.11 Power requirements for pumps.

BOX 7.2 EXAMPLE: CALCULATE ENERGY AND POWER REQUIRED TO FILL A TANK WITH WATER

A 600 m³ tank, 10 m high, is filled with water each day. Calculate the amount of energy required to fill the tank (Figure 7.12). If the pump runs for 6 h each day, calculate the power required.

First calculate the amount of energy needed each day using the energy formula

$$\text{energy (kWh)} = \frac{\text{volume (m}^3) \times \text{head (m)}}{367}.$$

Put in values for volume and head

$$\text{energy (kWh)} = \frac{600 \times 10}{367}$$

$$= 16.3 \text{ kWh.}$$

This is the energy required each day to fill the tank. Note that how the time period over which the energy is used needs to be specified.

Now calculate the water power needed. First calculate the discharge

$$\text{discharge (m}^3/\text{h)} = \frac{\text{volume (m}^3)}{\text{time (h)}}$$

$$= \frac{600}{6} = 100 \text{ m}^3/\text{h.}$$

Now calculate power

$$\text{power (kW)} = \frac{\text{discharge (m}^3/\text{h)} \times \text{head (m)}}{367}$$

$$= \frac{100 \times 10}{367} = 2.7 \text{ kW.}$$

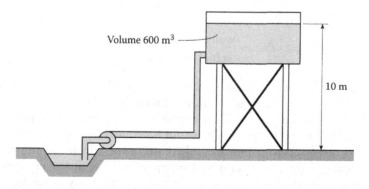

Volume 600 m³

10 m

Figure 7.12 Calculating energy and power to fill a tank.

The energy can also be calculated from power by multiplying power by the time over which the power is used. In this case, the time is 6 h.

energy (kWh) = water power (kW) × operating time (h)

= 2.7 × 6

= 16.3 kWh.

Note that the two approaches produce the same answer.

7.6.1 Efficiency

In the example in Box 7.2, the 'water energy' and 'water power' are only part of what is needed to actually fill the water tank. Losses occur in the power unit that drives the pump and in the pump itself, and these need to be taken into account when the total power is needed for pumping. Energy and power losses occur when fuel energy is converted into useful water energy and through turbulence and friction in the pump and fittings. Much of this is dissipated as heat energy. The losses that occur are expressed as an *efficiency*, which can be looked at from both a power and an energy point of view.

Energy and power efficiencies are often assumed to be the same. In practice, this may not be the case. A seasonal assessment of energy use efficiency may not always give the same value as power use efficiency measured only one or two times during a lengthy period of operation.

Power efficiency is a measure of how well the power from the power unit is converted into useful water power in the pump and is calculated as follows

$$\text{power efficiency (\%)} = \frac{\text{water power output}}{\text{actual power input}} \times 100.$$

Energy efficiency is used to judge how well a pump is performing over a longer period of time.

$$\text{energy efficiency (\%)} = \frac{\text{water energy output}}{\text{actual energy input}} \times 100.$$

To calculate energy efficiency, the time over which the energy is used must be known, for example, a day, month or a season.

A system with no energy losses would have an efficiency of 100% and so all the power or energy input would be transferred to the water. But this does not happen in practice. There are always losses in the various components of the power unit and pump as well as in the pipe system. The actual efficiency will be the product of the efficiency of each component. For example, a centrifugal pump with an efficiency of 80% being driven by an electric motor with an efficiency of 80% would have an overall efficiency of 64% $(0.8 \times 0.8 = 0.64)$. The efficiency of the pipe system can be incorporated in a similar way. So it is not just the efficiency of the pump that matters – all the components of the system must be well matched to produce a high overall efficiency. Just to complicate matters, the efficiency of each component is not fixed – the values vary depending on the discharge and pressure in the system. The designer's objective is to match the various components so that together they produce the highest level of overall efficiency at the desired pressure and discharge. Usually, this is not one fixed point but a range of pressures and discharges in which it is judged that the system is performing optimally with minimal energy losses.

This is true for all lifting devices and not just centrifugal pumps. Treadle pumps, for example, have an optimum treadling speed that minimises losses and power inputs from an operator and maximises pump output in terms of pressure and discharge (see Box 7.3).

BOX 7.3 EXAMPLE: WHAT DISCHARGE AND PRESSURE CAN A PERSON ACHIEVE WITH A TREADLE PUMP?

Estimate the discharge and pressures that can be expected from a treadle pump operated by a farmer in a developing country. Make reasonable assumptions to complete the calculations.

The first step is to determine the power available for pumping. Once this is established, it is possible to determine the pressure and discharge that could be expected from the pump.

Assume that the pump operator is a reasonably fit, well-fed male between 20 and 40 years. He could be expected to produce a steady power output of 75 W for long periods. To give you some idea of what this means, it is like walking up and down a flight of stairs in about 20 s. Not so bad for the first few times but think about sustaining this over several hours. In a developing country, where operators may not be so well fed, or operators are women and children, a more reasonable power value might be 30 W.

The pump unit is also not very efficient at converting human power into water power so we need to allow for an efficiency of say 40%.

So taking these figures into account, assume the power input is 30 W (0.03 kW) and the pump efficiency is 40%:

$$\text{water power (kW)} = \frac{\text{discharge (m}^3\text{/h)} \times \text{head (m)}}{367}$$

$$0.03 \times 0.4 = \frac{\text{discharge (m}^3\text{/h)} \times \text{head (m)}}{367}.$$

Power is directly related to both discharge and head, so at the same power input level, if head increases then discharge will decrease and vice versa.

So assume the water lift is 2 m which is typical for many treadle pump applications. Now calculate the discharge using the above power formula

$$\text{discharge} = \frac{0.03 \times 0.4 \times 367}{2} = 2.20 \text{ m}^3/\text{h}$$

$$= 0.6 \text{ L/s.}$$

If the water lift is reduced to 1 m, the discharge increases

$$\text{discharge} = 1.2 \text{ L/s.}$$

So when the pumping head is low, the discharge ranges from 0.6 to 1.2 L/s

In some countries, treadle pumps are used to pump water to much higher heads than this, but there is a limit to this which depends on the operator. The maximum pressure is determined by the weight of the operator who stands on the treadle and applies a downward force on the pistons.

So assume a typical 65 kg operator standing on a treadle immediately above a typical 100-mm diameter piston. Now calculate the pressure using

$$\text{Pressure (N/m}^2) = \frac{\text{force (N)}}{\text{area (m}^2)}.$$

Calculate the force and the area

$$\text{max. force from operator (N)} = \text{mass of operator (kg)} \times \text{gravity constant (m/s}^2)$$

$$= 65 \times 9.81$$

$$= 637 \text{ N}$$

$$\text{area of piston} = \frac{\pi \times d^2}{4} = \frac{\pi \times 0.1^2}{4} = 0.007 \text{ m}^2.$$

Put values for force and area into the formula to calculate pressure

$$\text{pressure} = \frac{637}{0.007} = 91,000 \text{ N/m}^2.$$

Calculate this in terms of head (m) using the pressure head equation

$$P = \rho g h$$

$$\text{head (m)} = \frac{P}{\rho g}$$

$$= \frac{91,000}{1,000 \times 9.81} = 9.27 \text{ m.}$$

So the maximum head this operator can achieve pressing his full weight onto the piston is 9 m. In practice, not all operators put their full weight on the treadles and so the actual pressure may be much lower than this, say 5 m. The pressure can be increased by reducing the diameter of the pistons but this will also reduce the volume of water lifted at each pump stroke. Another way of course would be to get a heavier operator or perhaps use two operators at the same time, which is often done when children are using it.

If the maximum pressure that can realistically be achieved on this pump is 5 m, then in such circumstances, the water power formula shows that the discharge would be reduced to 0.24 L/s.

Three important points arise from this:

- First, the pumping head is the sum of the suction lift and the delivery lift and so it is the pumping head that is 5 m. Treadle pump suppliers are notorious for quoting delivery and suction lifts separately and this can confuse the buyer who is trying to compare pump performance.
- Second, it is clear that you cannot have both high pressure and a large discharge. You can have one or the other and the combination depends on the power available from operator. Basically, there is no 'free lunch' – you get out of the pump what you are prepared to put into it.
- Finally, this is not rocket science. But it does demonstrate how basic hydraulics can be applied usefully to design what is a simple but very effective water lifting device – something that is not always done in practice.

7.7 ROTO-DYNAMIC PUMP PERFORMANCE

Small centrifugal pumps are sometimes characterised by the power of their drive motors, for example, 3 HP pump or a 5 kW pump, or by their delivery diameter, for example, 50 mm pump (Table 7.1). This provides some guidance for selection but the full performance characteristics should be used for larger pumps.

All roto-dynamic pumps are factory tested and data are published by the manufacturers on the following characteristics:

Table 7.1 Typical discharges from small centrifugal pumps

Pump size (mm)	Discharge (L/s)
25	0–5
50	5–15
75	15–25
100	25–35
125	35–50

- Discharge and head
- Discharge and power
- Discharge and efficiency

These data are usually presented graphically and typical characteristics for all three pump types are shown in Figure 7.13. The curves shown are

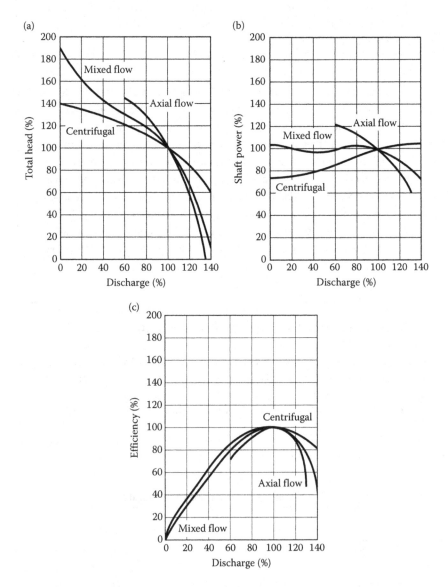

Figure 7.13 Pump characteristics. (a) Discharge-head. (b) Discharge-power. (c) Discharge-efficiency.

only for one operating speed. But as pumps can run at many different speeds, several graphs are needed to show their full performance possibilities.

7.7.1 Discharge and head

The relationship between discharge and head is usually of most immediate concern to the user. Will the pump deliver the discharge at the required pressure? A pump can, in fact, deliver a wide range of discharges but there will be changes in pressure as the discharge changes. Pump speed can also be varied and this changes both head and discharge. The faster the speed, the greater will be the head and discharge.

Figure 7.13a shows typical discharge-head curves for centrifugal, mixed flow and axial flow pumps for a given pump speed. The axes of the graph would normally show discharge (in m^3/s) and head (in m). But in order to show typical changes in performance, the curves have been drawn to show the percentage changes when either the discharge or the head are changed from the normal operating condition represented by the 100% point. So for the centrifugal pump, when the discharge is reduced to 80% of its design flow, the head increases to 112% of its design value. Note that these values of change are only representative values. They would be different for different pumps and so reference would need to be made to manufacturer's data.

Consider the centrifugal pump first. When it is started up and the delivery valve is closed, there is no discharge and the head is at its maximum. The graph shows that for this pump, the head reaches 140% of its design head. As the discharge increases, there is a trade-off between discharge and head. Less head is available when the discharge increases.

The discharge-head curves for both the mixed flow and axial flow pumps are similar in shape to that of the centrifugal pump. When the head is high, the discharge is low, and when the head is low, the discharge is high.

The curves for all three pumps are for a given speed of rotation. When the speed is changed, the curve will also change. The greater the speed, the greater will be the head and the discharge. So there will be several discharge-head curves for each pump, one for each speed.

7.7.2 Discharge and power

All pumps need power to rotate their impellers. The amount of power needed depends on the pump speed and the head and discharge required. For centrifugal pumps, the power requirement is low when starting up but it rises steadily as the discharge increases (Figure 7.13b). For axial flow pumps, the power requirement is quite different. There is a very large power demand when starting up because there is a lot of water and a heavy pump impeller to get moving. Once the pump is running, the power demand drops to its normal operating level. Mixed flow pumps operate in between these two contrasting conditions and have a more uniform power demand over the discharge range.

7.7.3 Discharge and efficiency

Power efficiency is a measure of the power input to the water power output and this varies over the operating range of all three pump types. It is low at low discharges but rises to a peak and then tails off. Pump selection involves choosing a pump pressure and discharge that occur when the pump is working at its maximum efficiency. This can typically be between 40% and 80%.

Generally, efficiency increases as the discharge increases. But it rises to some maximum value and then falls again over the remaining discharge range (see pump characteristics Figure 7.13c). The maximum efficiency is usually between 80% and 120% and so there is only a limited range of discharges and heads over which pumps operate at maximum efficiency. Outside this range, they will still work but they will be less efficient and so more power is needed to operate the same system. Smaller pumps are usually less efficient than larger ones because there is more friction to overcome relative to their size. But inefficiency is less important for small pumps.

7.8 CHOOSING THE RIGHT KIND OF PUMP

Specific speed (N_s) is one way of selecting the right kind of pump to use – centrifugal, mixed flow or axial. This is a number that depends on the speed of the pump, the discharge and head required for a particular installation and provides a common baseline for comparing pumps.

It is the speed at which a pump will deliver 1 m³/s at 1.0-m head and is calculated as follows

$$N_s = \frac{NQ^{1/2}}{H^{3/4}},$$

where N is rotational speed of the pump (rpm), Q is pump discharge (m³/s) and H is pumping head (m)

The specific speed is independent of the size of the pump and so it describes the shape of the pump rather than how big it is. But specific speed is not dimensionless. So it is important to make sure that SI units are used so that the range of specific speeds and pump types is as shown in Table 7.2.

Beware of specific speeds that seem to be quite different to those used here. The main reason is likely to be the use of different units of measurement. The United States still use feet and pounds as their basic units of measurement and so any specific speed they calculate will depend on these units.

Table 7.2 Specific speeds for different pumps

Pump type	Specific speed N_s	Comments
Centrifugal	10–70	High head – low discharge
Mixed flow	70–170	Medium head – medium discharge
Axial flow	Above 170	Low head – large discharge

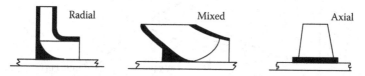

Figure 7.14 Typical pump impeller shapes.

The shape of a pump impeller is closely related to its speed of rotation and the pumping head. These factors are combined into a number known as the *specific speed* which provides a common base for comparing pumps.

The shape of the pump impeller helps to define the different pump types (Figure 7.14). Centrifugal pumps produce high pressures using centrifugal force. To achieve this, the impeller is shaped to turn the flow from the pump inlet through a right angle so that it moves radially outwards towards the delivery pipe as the impeller spins (left-hand picture Figure 7.14). It is this radial flow and the large ratio between the pump inlet diameter and the outlet diameter that generates high velocities, and hence the high pressures associate with centrifugal pumps. In contrast, axial flow pumps produce large discharges rather than high heads and so no centrifugal forces are needed. The impeller is propeller shaped and is designed to move large volumes of water along the axis of the pump. It works in much the same way as a propeller pulls an aircraft through the air (right-hand picture Figure 7.14). Note that the ratio of the inlet diameter to the outlet diameter is 1.0. In between these two extremes are mixed flow pumps. These variously have a mixture of radial flow and a larger outlet diameter – to produce some pressure, and axial flow – to produce flow.

7.9 MATCHING A CENTRIFUGAL PUMP WITH A PIPELINE

Selecting the right centrifugal pump and pipe system for a particular water supply installation depends both on hydraulics and cost. There is not just one unique solution to the problem. There will be several pump and pipeline combinations that will do the job from a hydraulic point of view but the final choice usually comes down to cost – which combination is the cheapest? Will a small diameter pipe with a large pump to overcome the

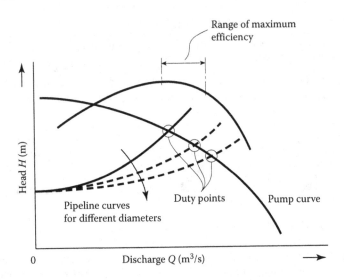

Figure 7.15 Matching a pump with a pipeline.

high head losses be more or less costly than a larger and more expensive pipeline with a small pump that is less expensive to run? Working out the capital costs of each pump and pipeline together with the fuel and maintenance costs over the lifetime of the installation will provide the answer.

First, the hydraulics. An example of matching a pump and pipeline is shown Figure 7.15. The pump is represented by two curves – a discharge-head curve and a discharge-efficiency curve. The pipeline is represented by a head-discharge curve which shows how head losses in the pipe change when it is carrying a range of different discharges. This is constructed by putting discharge values into a head loss formula, such as the Darcy–Weisbach formula, and calculating the resulting head loss. The corresponding discharge and head loss values are then plotted onto the graph. Note that the pipeline curve does not start at zero on the head axis but at some point which represents the total static head on the pump. This will be the elevation change from the water level in the sump to the point of delivery. This will be a fixed value for a given installation and not dependent on discharge. The intersection of the pump curve and the pipeline curve gives the pressure and discharge at which the combined pump and pipeline will operate. If this intersection occurs within the acceptable efficiency range for the pump, the combination is acceptable from a hydraulic point of view. A practical example of how to match a pump with a pipeline is shown in Box 7.4.

Next comes the cost bit. The hydraulic example above is for one pipeline and pump combination. But other pipe diameters could do the job just as well. Figure 7.15 shows how several pipelines of different diameters can be assessed each producing a different duty point. More pump curves could also be introduced to see how each pipeline would interact with a range of different pumps. The end result is a wide range of possible pipeline and pump combinations that can do the job. But some will be cheaper than others and the task then is to find out which is the cheapest. This is done by comparing not just the capital costs of each pipeline and pump but a combination of both the capital costs and the running costs (energy and maintenance) over the economic life of the system. An example of how to do this is shown in Box 7.5.

BOX 7.4   EXAMPLE: MATCHING A PUMP WITH A PIPELINE

Water is pumped from a river through a 150 mm diameter pipeline 950 m long to an open storage tank with a water level 45 m above the river. A pump is available and has the discharge-head performance characteristics shown below. Calculate the duty point for the pump when the friction factor for the pipeline $\lambda = 0.04$.

Total head (m)	30	50	65	80
Discharge (L/min)	2,000	1,750	1,410	800

The first step is to plot a graph of the pump discharge-head characteristic (Figure 7.16).

Next, calculate the head loss in the pipeline for a range of discharge values using the Darcy–Weisbach formula

$$h_f = \frac{\lambda l v^2}{2gd},$$

when $\lambda = 0.04$, length $l = 950$ m and diameter $d = 0.15$ m

$$h_f = \frac{0.04 \times 950 \times v^2}{2 \times 9.81 \times 0.15} = 12.9\,v^2.$$

But from the discharge equation

$$v = \frac{Q}{a},$$

and so

$$v^2 = \frac{Q^2}{a^2}.$$

Calculate area a

$$a = \frac{\pi d^2}{4} = \frac{\pi \times 0.15^2}{4} = 0.017\ \text{m}^2.$$

Use these values to calculate h_f

$$h_f = 12.9 \times \frac{Q^2}{0.017^2}$$
$$= 44{,}636\,Q^2.$$

Now calculate values of head loss for different values of Q. Remember the values of discharge need to be in SI units of m^3/s for the calculation. Also the static head of 45 m must be added to the calculated head loss to determine the total pumping head. The results are tabulated below:

Discharge (L/min)	Discharge (m³/s)	Head loss h_f (m)	Head loss + 45 m (m)
2,000	0.33	48.6	93.6
1,600	0.027	30.17	75.15
1,200	0.02	17.85	62.85
1,000	0.017	12.30	57.30
500	0.006	1.60	46.60

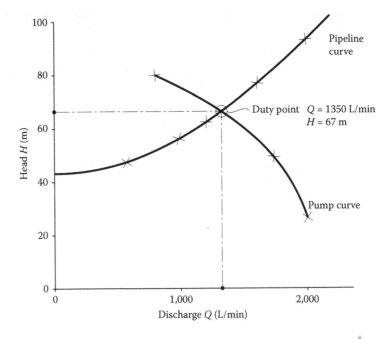

Figure 7.16 Matching a pump with a pipeline.

Now construct a graph of the pump curve and the pipe curve (Figure 7.16). From the graph, the duty point is located where the two curves cross. This occurs when

$Q = 1350$ L/min,

and

$H = 67$ m.

BOX 7.5   EXAMPLE: SELECTING THE CHEAPEST PIPELINE AND PUMP COMBINATION

A pumped water supply pipeline, 1,150 m long, is to deliver 2,700 L/min. The static lift is 20 m, the friction factor for the pipeline $\lambda = 0.03$ and the pumping plant operates at 80% efficiency. If the system runs for 17 h each day, 365 days per year, determine the cheapest combination of pipeline and pump for this installation given the following data:

Item	Cost	Economic life (years)	Maintenance costs
Pumps		20	10% per annum
One kilowatt of installed power	£150		
Electricity	1.8 p/kWh		
Pipelines			0%
100 mm	£5.90/m	50	
150 mm	£8.00/m	50	
200 mm	£10.50/m	50	
250 mm	£14.80/m	50	
300 mm	£21/m	50	

Interest rate 5%

Now calculate the power for each of the pipe and pump combinations. The results are tabulated below.

To calculate the cost of pumping, first determine the power requirement of each pipe–pump combination. So use the power formula

$$\text{power (kW)} = \frac{\text{discharge (m}^3\text{/h)} \times \text{head (m)}}{367 \times \text{efficiency}}.$$

We know the discharge is 2,700 L/min (162 m³/h) and the efficiency is 80% (0.8 in the formula) but we now know the head which is a combination of the static head and the head loss in the pipeline. So first calculate the head loss h_f using the Darcy–Weisbach formula:

$$h_f = \frac{\lambda l v^2}{2gd}.$$

To calculate velocity use the discharge equation

$$v = \frac{Q}{a}.$$

Calculate area a

$$a = \frac{\pi d^2}{4}.$$

Put this into the equation for v

$$v = \frac{4Q}{\pi d^2}.$$

Pipe diameter (mm)	Velocity (m/s)	Head loss (m)	Static head	Total pumping head (m)	Power (kW)
100	5.73	577.84	20	597.84	329.89
150	2.55	76.09	20	96.09	53.03
200	1.43	18.06	20	38.06	21.00
250	0.92	5.92	20	25.92	14.30
300	0.64	2.38	20	22.38	12.35

In order to combine and compare capital costs and operating costs of the various pipeline–pump combinations, calculate the equivalent annual costs for each system (if you are not familiar with discounting cash flows, then please refer to any basic economics text book).

First calculate the annual pump cost

annual pump cost (£) = energy cost + annual replacement cost
+ maintenance cost + annual interest cost.

Calculate each component of the annual pump cost

energy cost (£) = operating time (h) × power (kW) × cost (£/kWh)
= 17 × 365 × power (kW) × cost (£/kWh)
= 6,205 × power (kW) × cost (£/kWh).

The pump must be replaced every 20 years so the annual replacement cost is 5% of the capital cost of the pump.

annual replacement cost = %annual replacement × capital cost (£)
= 0.05 × £150 × installed power (kW).

maintenancecost = 10%of capital cost
= 0.1 × £150 × installed power (kW).

annual interest cost = pump capital cost (£) × interest rate (%)
= £150 × installed power (kW) × interest rate (%).

Add all these costs together for each pump to obtain the annual cost of pumping:

Pipe diameter (mm)(1)	Capital cost (£)(2)	Energy cost (£) (3)	Replacement cost (£) (4)	Interest cost (£) (5)	Maintenance cost (£) (6)	Total annual cost (£) (3)+(4)+(5)+(6)
100	49,484	36,845	2,474	2,474	4,948	46,741
150	7,953	5,922	397	397	795	7,511
200	3,150	2,345	157	157	315	2,974
250	2,145	1,597	107	107	214	2,025
300	1,852	1,379	92	92	185	1,748

Now calculate the annual cost of each pipeline over the economic life of the pipeline (note there is no maintenance cost)

annual pipe cost (£) = annual replacement cost + annual interest cost.

Calculate each component of the annual pipeline cost
The pipe must be replaced every 50 years so the annual replacement cost is 2% of the capital cost of the pipe.

annual replacement cost = %annual replacement × capital cost(£)
= 0.02 × capital cost (£).

annual interest cost = capital cost (£) × interest rate (%).

Add all these together to determine the annual cost of the pipelines:

Pipe diameter (mm)	Capital cost (£) (1)	Replacement cost (£) (2)	Interest cost (£) (3)	Total annual cost (£) (2)+(3)
100	6,785	135	339	474
150	9,200	184	460	644
200	12,075	241	603	845
250	17,020	340	851	1,191
300	24,150	483	1,207	1,690

Now add the annual pumping costs to the annual pipeline costs to obtain the total cost for each pump and pipeline combination:

Pipe diameter (mm)	Pump costs (£) (1)	Pipeline costs (£) (2)	Combined cost (£) (1) + (2)
100	46,741	474	47,215
150	7,511	644	8,155
200	2,974	845	3,819
250	2,025	1,191	3,216
300	1,748	1,690	3,438

The costs show that the 250 mm diameter pipe is the lowest cost solution.

7.10 CONNECTING PUMPS IN SERIES AND IN PARALLEL

There are many situations when one centrifugal pump is not enough to deliver the required head or discharge and so two or more pumps are needed. There may also be circumstances when discharge requirements vary widely, such as in meeting domestic water demand, and it is preferable to have several small pumps working together instead of a large one. Centrifugal pumps can be operated together either in *series* or in *parallel* (Figure 7.17).

Pumps are connected in series when extra head is required. Note that the pumps need to be identical. They are connected together with the same suction and delivery pipe but are powered by different motors. The same flow passes through pump 1 and then through pump 2 and so the discharge is the same as it would be for one pump but the head is doubled. The discharge-head curve for two pumps is obtained by taking the curve for one pump and doubling the head for each value of discharge.

Pumps are operated in parallel when more discharge is required. Again the pumps must be identical. They each have separate suctions but they are connected into a common delivery pipe. With this type of connection, the head is the same as for a single pump but the discharge is doubled. The discharge-head curve for the two pumps is obtained by taking the curve for one pump and doubling the discharge for each value of head.

But doubling the discharge on the pump curve does not mean that the system discharge is also doubled. The new combined pump curve must be matched with the pipe curve to determine the new discharge and pressure at which the new setup will work. The same is true when the pump head is doubled in series connections.

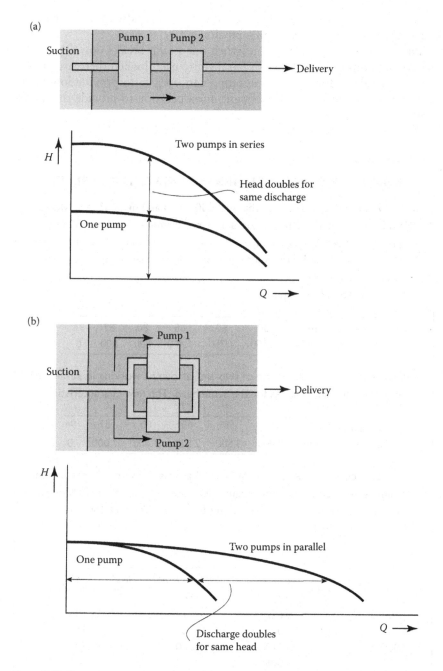

Figure 7.17 Pumps in series and in parallel. (a) Series. (b) Parallel.

An example in Box 7.6 illustrates how two pumps working in parallel affect the discharge and head in a pipeline.

Pumps in series work well, but when one pump breaks down, then the whole system is down. Pumps in parallel are useful when there are widely varying demands for water. One pump can operate to provide low flows and the second pump can be brought into operation to provide the larger flows.

BOX 7.6   EXAMPLE: PUMPS WORKING IN PARALLEL

In Box 7.4, the duty point – discharge 1,350 L/min and head 67 m – was calculated for a centrifugal pump and a 150 mm diameter pipeline 950 m long supplying an open storage tank with a water level 45 m above the river. Determine the revised duty point when a second similar pump is connected in parallel into the system.

Characteristic of single pump:

Total head (m)	30	50	65	80	87	94
Discharge (L/min)	2,000	1,750	1,410	800	500	0

Determine the characteristic of two similar pumps working in parallel by doubling the discharge values for the same head. The new characteristic is:

Total head (m)	30	50	65	80	87	94
Discharge (L/min)	4,000	3,500	2,820	1,600	1,000	0

The pipe curve is the same as in Box 7.4 as this does not change but the data needs extending to take the higher discharges into account. The formula for calculating the head loss is based on the Darcy–Weisbach formula

$$h_f = 44,636\, Q^2.$$

Discharge (L/min)	Discharge (m³/s)	Head loss h_f (m)	Head loss + 45 m (m)
2,000	0.33	48.6	93.6
1,600	0.027	30.17	75.15
1,200	0.02	17.85	62.85
1,000	0.017	12.30	57.30
500	0.006	1.60	46.60

Now construct a graph of the pump and pipe curves (Figure 7.18).

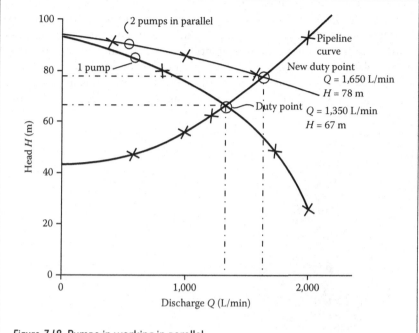

Figure 7.18 Pumps in working in parallel.

From the graph, the duty point is located were the two curves cross.
This occurs when $Q = 1\,650$ L/min and $H = 78$ m.

7.11 VARIABLE SPEED PUMPS

Centrifugal pump installations are designed to the meet the maximum discharge at a particular running speed. Lower discharges are then dealt with by throttling the flow using various types of valve installed along the pipeline. This works fine as a control mechanism but the pump is still using up energy as if it was delivering the design discharge and so a lot of energy is wasted. A typical electric motor costing £500 could consume over £50,000 in electricity charges over its useful life and so a single percentage point increase in efficiency can save as much as the capital cost of the motor itself.

One solution to the varying flow problem is to vary the pump speed to fit with the change in demand. But this is not always convenient and easy to do. Electric motors, which are increasingly used to drive pumps, are usually AC induction motors and they run at a fixed speed which is determined by the frequency of the alternating electricity supply. In the UK, this is 50 Hz (50 cycles per second) so motors (and pumps) run at 3,000 rpm. In some countries, the AC supply is 60 Hz and so motors there run at 3,600 rpm.

It is possible to vary motor speed by setting a number of predetermined speeds using different motor wiring systems. But another way is to vary the frequency of the electricity supply. This is a recent innovation for controlling pumps. The pump motor is fitted with a frequency converter and various electronic sensors so that the pump can be set to run at a constant pressure over a wide range of discharges or at a constant discharge over a wide range of pressures (Figure 7.19). They are often referred to as

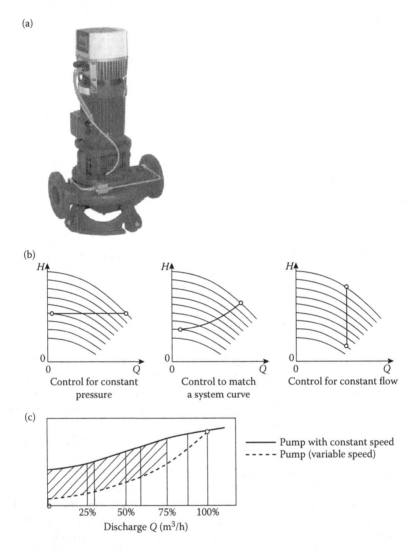

Figure 7.19 Variable speed pumps. (a) Pump fitted with electric motor power unit and frequency converter. (b) Pump output matched to demand by varying motor speed. (c) Difference in power consumption for fixed and variable speed pumps.

inverters but this describes only part of their function. The frequency converter adjusts the motor and hence the pump speed so that the various set demands are met. This can significantly reduce energy demands and the costs of pumping.

7.12 OPERATING PUMPS

Here are some good practice guidelines for operating pumps properly.

7.12.1 Centrifugal pumps

Centrifugal pumps always have control valves fitted on the suction and delivery sides of the pump. The valve on the delivery side is used to control pressure and discharge (Figure 7.8). It is closed before starting so that the pump can be primed. Once the pump is running, it is slowly opened to deliver the flow.

A reflux (non-return) valve is also usually fitted after the delivery valve to stop reverse flows that can damage the pump (see Section 7.14). Some reflux valves have a small by-pass valve fitted which allows water stored in the pipe to pass around the valve and be used for priming the pump.

A reflux valve is usually fitted on the end of the suction pipe. This keeps the suction pipe full of water and makes priming easier.

Before starting a pump, the delivery valve must be closed. The pump and the suction pipe are then filled with water for priming. When pumps are located below the sump water level, they are primed automatically as water flow under gravity into the pump casing. Contrary to expectation, starting a pump with a closed delivery valve does not cause problems. The pressure does not go on rising and damage the pump. Rather it reaches a steady speed and pressure. Remember the pressure depends only on the speed of rotation of the impeller. The faster it rotates, the higher will be the pressure. The delivery valve is then slowly opened, water flows into the pipeline, and as the discharge increases, the pressure at the pump gradually falls. At the same time, the power requirement increases. This gradual change continues until the delivery pipe is full of water and the system is operating at its design head and discharge.

These operating procedures are important for several reasons. Suppose for example, the delivery valve was left open as often happens. Not only would it be difficult to prime the pump but also when it was started, water would surge along the delivery pipe. Such a sudden surge of water might damage valves and fittings along the pipeline. The rapid increase in discharge also causes a rapid increase in power demand and power units do not usually like this. It is like getting into a car when the engine is cold and suddenly trying to put it into top gear and accelerate fast. The power demand is too great and the engine will stall and stop. The most sensible way to get moving is to build up the power demand gradually through the use of the gears and clutch. The same idea applies to pumps by opening the delivery valve slowly.

7.12.2 Axial flow pumps

Axial flow pumps do not have any control valves on either the suction or the delivery. Because of the high power demand when starting up, it is not desirable to start a pump against a closed valve. What is needed, however, on some pumps is a siphon breaker. This stops water from siphoning back from the delivery side into the sump when the pump stops by allowing air into the pump casing.

7.13 POWER UNITS

There are two main types of power unit:

- Internal combustion engines
- Electric motors

7.13.1 Internal combustion engines

Many pumping installations do not have easy access to electricity and so rely on petrol (spark ignition) engines or diesel (compression ignition) engines to drive pumps. These engines have a good weight to power output ratio, and are compact in size and relatively cheap due to mass production techniques.

Diesel engines tend to be heavier and more robust than petrol engines but are more expensive to buy. However, they are also more efficient to run, and if operated and maintained properly, they have a longer working life and are more reliable than petrol. In some countries, petrol-driven pumps need replacement after only 3 years' operation. Diesel pumps operating in similar conditions could be expected to last at least 6 years. However, it must not be forgotten that engine life is not just measured in years, it is measured in hours of operation and its useful life depends on how well it is operated and serviced. There are cases in developing countries where diesel pumps have been in continuous use for 30 years and more. A diesel pump can be up to four times as heavy as an equivalent petrol pump, and so if portability is important, a petrol engine may be the answer.

7.13.2 Electric motors

Electric motors are very efficient in energy use (75%–85%) and can be used to drive all sizes and types of pumps. The main drawback is the reliance on a power supply which is outside the control of the pump operator, and in some countries, it is unreliable. Inevitably, electrical power supplies usually fail when they are most needed and so backup generators may be needed driven by diesel engines. Some people are now using solar panels and wind turbines to provide the electricity.

7.14 SURGE IN PUMPING MAINS

Water hammer in pipelines has already been discussed in Section 4.14 but it can cause particular problems in pumping mains by creating *surges*. This is the slower mass movement of water often resulting from the faster moving water hammer shock waves.

In pumping mains, the process is reversed and it is the surge of water that can set up the water hammer. To see how this happens, imagine water being pumped along a pipeline when suddenly the pump stops (Figure 7.20). The flow in the pipe does not stop immediately but continues to move along the pipe. But as the main driving force has now gone, the flow gradually slows down because of friction. As the flow moves away from the pump and as no water enters the pipe, an empty space forms near the pump. This is called *water column separation* and the pressure in the empty space drops rapidly to the vapour pressure of water.

The flow gradually slows down through friction and stops. But there is now an empty space in the pipe and so water starts to flow back towards the pump gathering speed as it goes. It rapidly refills the void and then comes to a sudden and violent stop as it hits the pump. This is very similar to suddenly closing a valve on a pipeline and results in a high-pressure shock wave which moves up the pipe at high speed (approximately 1,000 m/s). This can not only burst the pipe but may also seriously damage the pump.

There are several ways of avoiding this problem:

Stop pumps slowly: When pumps stop slowly, water continues to enter the pipe and so the water column does not separate and water hammer is avoided. Electric pumps are a problem because they stop very quickly when the power fails. Diesel pumps take some time to slow down when the fuel is switched off and this is usually enough time to avoid the problem.

Use a non-return valve: Using a non-return valve in the delivery pipeline will allow the flow to pass normally along the pipe but stop water from flowing back towards the pump. In this case, the valve and the pipe absorb the water hammer pressures and the pump is protected.

Use an air vessel: This is similar in action to a surge tank but in reverse (Figure 7.20b). When a pump stops and the pressure starts to drop, water flows from a pressurised tank into the pipeline to fill the void and stop the water column from separating. When the flow stops and

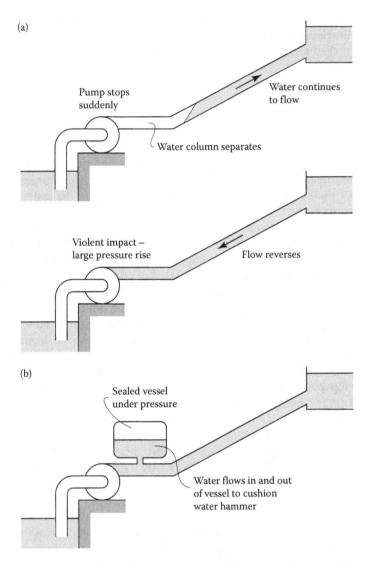

Figure 7.20 Water hammer in pumping mains. (a) Effects of suddenly stopping a pump. (b) Air vessel used to reduce surge in pumping mains.

reverses, it flows back into the tank. The water then oscillates back and forth until it eventually stops through friction. Unlike a surge tank, an air vessel is sealed and the air trapped inside acts like a large coiled spring compressing and expanding to dampen the movement of the water in much the same way as a shock absorber on a car dampens the large bumps in the road. This device tends to be expensive and so would only be used on large, important pumping installations where the designers expect serious water hammer problems to occur.

7.15 TURBINES

Water can also produce energy as well as absorb it. This idea has been exploited for centuries long before rotary pumps where invented. Water wheels were used by the Romans and throughout Europe to grind cereals. Today these wheels have been replaced by turbines that are connected to generators to produce electrical energy. These are used extensively in Scotland for power generation, in mountainous countries such as Switzerland and in many other countries where convenient dam sites can be located to produce the required head. Unfortunately, there are not many sites around the world where there are sufficient renewable energy sources that can be exploited in this way. They are not usually the main source of electricity but provide a valuable addition to the main power source, such as coal, gas or oil-fired power stations, to meet peak demands.

Another development has been the combination of pumps and turbines in what is known as a pumped storage scheme. In Wales, for example, a small hydropower plant is used to generate electricity during the day to meet peak demands. During the night, when power demand is low, surplus electricity from the main grid is used to pump water back up into the reservoir to generate electricity the next day. This is a way of 'storing' electricity by storing water.

There are three main types of turbine (Figure 7.21):

- Impulse turbines
- Reaction turbines
- Axial flow turbines

Impulse turbines are similar in operation to the old water wheel. The most common is the Pelton wheel, named after its American inventor, Lester Pelton (1829–1908). This comprises a wheel with specially shaped buckets around its periphery known as the runner. Water from a high-level reservoir is directed along a pipe and through a nozzle to produce a high-speed water jet. This is directed at the buckets and causes the runner to rotate. The momentum change of the jet as it hits the moving buckets creates the force for rotation. So by knowing the speed of the jet, it is possible to work out this force and the amount of electrical energy that can be generated. Pelton wheels can be very efficient at transferring energy from water to electricity and figures as high as 90% are quoted by manufacturers. They are best suited to high heads above 150 m. Some installations run with heads in excess of 600 m

Reaction turbines are like centrifugal pumps in reverse. The most common design is the Francis turbine. Although James Francis (1815–1892) did not invent the turbine, he did a great deal to develop the inlet guide vanes and runner blades to improve turbine efficiency and so his name was associated with it. The turbine resembles a centrifugal pump but instead of the

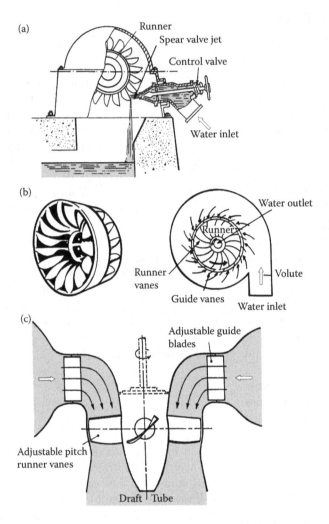

Figure 7.21 Turbines. (a) Pelton wheel. (b) Francis turbine. (c) Kaplan turbine.

rotating impeller driving the water as in a pump, the runner is driven by the water. Water is fed under pressure and at high velocity around a spiral casing onto the runner, and as it rotates, the pressure and kinetic energy of the water are transferred to the runner. Francis turbines normally work at heads of 15–150 m.

Axial flow (or propeller) turbines are known as Kaplan turbines. They are named after Victor Kaplan (1876–1934), a German professor who developed this type of turbine with adjustable blades. Axial flow turbines operate like axial flow pumps in reverse. They operate at low heads (less than 30 m) and are used in tidal power installations. They have blades on their runners that can be twisted to different angles in order to work at high efficiency over a wide range of operating conditions. This kind of turbine is also used in pumped storage schemes. The turbines are used to generate electricity throughout the night to pump water into reservoirs when the demand for electricity from the national grid is low. During the day, when the demand for electricity is high, water is released from the reservoir to drive the turbines and generate electricity for the grid.

7.16 SOME EXAMPLES TO TEST YOUR UNDERSTANDING

1. A centrifugal pump is connected to a 25 m length of 300 mm diameter suction pipe and discharges into a 500-m long pipe with a diameter of 150 mm. The water level in the river is 2 m above datum, the centre line of the pump is 3.75 m higher and the discharge pipe enters a tank whose water level is 25 m above datum. Calculate the pressure head that the pump must produce to pump a discharge of 0.150 m³/s. Assume a friction factor $\lambda = 0.015$ (61.8 m).

2. A pump supplies water to a large reservoir through a 2,000-m long pipeline 400 mm in diameter. If the difference in water level between the sump and the reservoir is 20 m and the friction factor of the pipeline is $\lambda = 0.03$, calculate the pressure and power output required to deliver a discharge of 0.35 m³/s (29.38 m; 101 kW).

3. A centrifugal pump lifts water from a well to a storage tank with an outlet 20 m above the water level in the well. The suction and delivery pipes are 50 mm in diameter and the total pipe length is 150 m. The friction factor for the pipeline is $\lambda = 0.035$. The pump performance at different heads is tabulated below:

Head (m)	0	12	25	30	32	33
Discharge (L/s)	9	8	6	4	2	0
Efficiency	0	50	60	60	50	0

Plot the pump characteristic curves and the pipe system curve and determine the discharge, head, efficiency and power required at the duty point (2.8 L/s; 31 m; 56%; 0.85 kW). If a second pump is connected in series calculate the new duty point (4.8 L/s; 57 m; 60%; 2.64 kW).

Chapter 8

Water resources engineering

8.1 HARNESSING THE FORCES OF NATURE

Civil engineering is about harnessing the forces of nature for the benefit of mankind, and this applies particularly to water. Engineers use the principles and practices of hydraulics to plan, design, and build hydraulic infrastructure, such as water supply and wastewater treatment systems, irrigation and drainage schemes, and hydro-electric schemes. They each include many of the elements described in this book, such as pipes and canals, dams, barrages, measuring weirs and flumes, and pumping stations. All these elements enable society to monitor, control, and manage water resources.

Although one water supply system can look very much like another to the untrained eye, no two systems are ever quite the same. Each new scheme brings with it a whole new set of problems to solve. The civil engineer's job is to use the same basic principles and apply them to the different physical and socio-economic circumstances and come up with appropriate and innovative solutions. There are often several solutions to the same problem and so engineers, working with other disciplines, will develop a range of feasible technical options to choose from. The final choice will be influenced by the costs and benefits associated with each solution. The views of the community and local politics will play an important part, as will the project's general appearance. Good design can make even the humblest pumping station attractive and add value to the landscape (see Box 8.1).

8.2 MAKING CHOICES IN A COMPLEX WORLD

Engineers are always faced with lots of choices. Choices about form and function, about size and shape, materials, quality and reliability, access and security, operation and maintenance and the cost implications of different options. This includes both the cost of constructing the works (capital costs) and operation and maintenance (recurrent costs). So engineers grow up in a culture of negotiation, trade-offs and compromise.

BOX 8.1   ST GERMANS PUMPING STATION: FITTING INFRASTRUCTURE WITH LOCAL CIRCUMSTANCES

St Germans drainage pumping station in Norfolk, UK, was commissioned in 2011 and is one of the largest in Europe. This well-designed low-profile building fits neatly into the countryside (Figure 8.1). The station protects highly productive agricultural land worth £3.5 billon and 25,000 homes, all lying below sea level. This £38 million station is designed not to fail. It is packed with the latest technology for monitoring and controlling discharges into the North Sea. All the pumps are set above high water levels in case the sea breaches the local defences. Although it is all electric, there are diesel generators built into the design in case the electricity supply fails. Enough fuel is stored on site to keep all the pumps running at full discharge for 10 days lifting 20 tons of water every second.

Figure 8.1 St Germans land drainage pumping station in Norfolk, UK.

Over the past 50 years, engineers have seen many changes to their role. Prior to the 1960s, and not long after the Second World War, the challenge in the United Kingdom (UK) was to regulate water resources to benefit people. The words 'environment', 'water pollution' and 'ecology', were far from being household words. People wanted reliable services and trusted engineers to make the choices and get on with doing their job. Thus, designing and building hydraulic infrastructure was a relatively straight-forward technical issue. But in more recent times we have seen growing awareness and concerns about our aquatic environment and so the challenge has shifted towards regulating people for the sake of our water resources. The public view of what engineers do has also changed to the point that some think they are working towards systematically destroying and polluting our environment in order to achieve material and economic growth (see Box 8.2).

In the poorer developing countries, it is not surprising that development is following a similar pathway. Poverty is the major issue across much of Africa and Asia with over 1 billion people still living on less than US$1.25 a day. In such circumstances, improving living standards takes precedence over the environment. Development is technology led and investment is very much about providing clean, reliable water supplies and sanitation, building irrigation schemes to increase food security and hydro schemes to generate much needed energy. At the moment, people come first, but international pressure is growing to change this approach as various agencies use their funds and loans to change attitudes.

BOX 8.2   CIVIL ENGINEER'S STORY

Civil engineers play a key role in complex infrastructure development and as such they are often seen as the people who inflict 'unsightly' structures on society rather than accepting that the blame for the misuse of technology lies with all of society. We tend to get what we wish and are willing to pay for. This is not a new story as engineers throughout history have suffered from the 'blame game', like the messenger bearing unwelcome news. In ancient Greece, engineers did not fare any better in spite of the benefits they brought. In Plato's Gorgias (380 BC), Socrates, in defence of engineers, says (paraphrased) *'you would not think to put the engineer on the same level as the lawyer and nor would you allow him to marry your daughter. But the engineer is not in the habit of magnifying his importance and like the army general or the ship's captain he has the ability to save lives: he can even save whole cities'.*

Civil engineers' reputation seems to have changed little, except for a brief spell in Victorian times when engineers like Isambard Kingdom Brunel and others were revered for their great engineering works and contribution to the industrial revolution. Today, engineers do not often come into direct contact with the public, indeed many seem to avoid the limelight. They also have an unenviable 'boiler-suit and spanners' image which seems almost impossible to shake off, at least in the UK. Professional engineers need to devote much more time and energy to improving communications with the public so that their important role in sustainable economic growth and protecting the environment is better recognised. Perhaps their reputation will rise someday to rank with architects, lawyers and doctors – and also their salaries!

In the richer countries, planning, designing and building hydraulic works has become a whole lot more complicated. People want the benefits of modern engineering such as clean water, cheap energy and food on their plates. But equally they are much more aware of the impact of engineering works on society, life quality and our natural environment. We still need to build new water infrastructure and maintain existing systems to meet the demands of growing populations. But there is now a strong desire for those affected by the work to participate and influence decision-making. Civil engineers are usually at the centre of the debate as they will eventually design and build the agreed solution. During the planning process they will be involved in negotiating a whole range of often conflicting physical, environmental and social requirements. But finding that compromise solution has become more and more complex and time consuming.

Even archaeology is now part of the process. When farmers are building irrigation reservoirs they are obliged to make sure that they will not be disturbing ancient sites. An archaeological investigation for a £1 million reservoir can add as much as £80,000 to the overall cost for which the farmer must pay.

Across Europe, where there are designated Sites of Special Scientific Interest (SSSIs), discussions with many and varied interest groups on types of structures and their location can often delay construction for many years. In the Thames valley in UK, where engineers are looking into the possibility of a new major storage reservoir for London, it is estimated that, once a decision to build is made in principle, it could take up to 20 years before it comes on-stream because of all the interests, conflicts and compromises to be negotiated.

The following is a brief look into how civil engineers engage in developing water supplies, wastewater treatment, irrigation and in protecting rivers and the aquatic environment.

8.3 WATER SUPPLY ENGINEERING

Water supply engineering is mostly technical common sense. It involves estimating the quantity of water required and then finding, storing, treating and delivering it to customers. Water storage is based on estimates of future population, the average water use/person/day, and the likelihood of new industries being set up. The treatment works and pipe network will need to cope with the likely maximum daily discharge, such as peak flow at the worst hour on a hot day. Added to this will be a discharge contingency for emergencies such as firefighting.

A key question for the engineers is how far ahead should schemes be planned? If it is 50 years then much investment is tied up for 30–40 years in oversized treatment works, pipes and pumps. So compromises are needed looking at several design periods. Dams, tunnels and other infrastructure that would be costly to enlarge are likely to be designed to meet requirements over the next 50 years. Whereas pipelines, treatment works and pumps, which can be extended or replaced more easily, would be designed for much shorter time periods at lower immediate cost.

Predicting future demand is not easy because of unexpected changes in population numbers and water use, housing densities and rates of industrial development. But growth projections are needed as the basis for designing the system. The average water use in the UK is about 150 L/person/day. In the United States it is over 500 L/person/day, and in much of Africa it is less than 20 L/person/day. Winter use may be 20% lower than the average and summer use 20% higher. In places were supplies are metered, people become more aware of the amount of water they are using and paying for and so they tend to use less. The downside of this is when people cut down too much and conditions in the home become insanitary.

8.3.1 Typical water supply scheme

Figure 8.2 shows a typical water supply system for a town. Water storage is needed to balance the variations in supply from rivers with the demand requirements of the town. A dam built across the river provides on-stream storage. The hatched line shows the extent of the water basin which drains into the river. An alternative is to construct an off-stream reservoir alongside the river. People demand a high level of security for water supplies and so storage must be designed to ensure there is always enough water

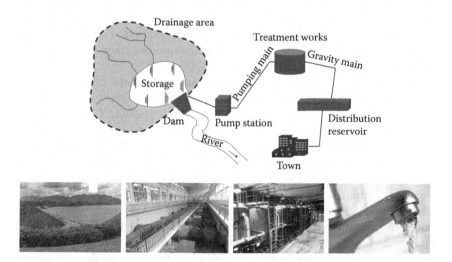

Figure 8.2 Water supply system for a town from dam to tap.

even in extreme dry years. Keeping the storage at acceptable levels will depend on the volume of water being used and on the volume flowing in from the basin. This may be continuous or intermittent depending on the climate. When rainfall and river flows are regular and reliable, the storage required may only need to be 6 months supply. But when flows are less reliable, particularly during drought periods, the required storage may exceed 2 years supply or more. Engineers designing reservoirs will take account of evaporation and seepage losses and also the amount of water that must be released to satisfy downstream water rights and environmental needs. Allowing the river just to run dry will be unacceptable but this may well happen in periods of severe drought when priority is given to conserving crucial drinking water supplies. Groundwater is another supply option if it is available in the right quantities. This is natural storage which costs nothing but the problem is predicting the amount you can safely abstract without damaging the aquifer.

The pumping station downstream of the dam and pressurised pipeline provide a steady discharge to the treatment works. A gravity pipeline takes treated water to a distribution reservoir close to the town. This is usually covered to keep water clean. The reservoir evens out the fluctuations in demand throughout the day. It also provides emergency and contingency supplies in case the pumping station or the treatments works are closed because of breakdown or maintenance. Water then flows through a pipe network to individual households and businesses. Most households have a similar (but smaller) distribution tank in their roof space to even out the fluctuations in demand within the household.

After the water is used, the wastewater flows out from households and industries through a treatment facility and back into the river. This is common practice in many countries to dilute treated wastewater with natural river water as it helps to further clean up the pollution. This is important because the combined flow may well become another town's water supply further downstream (see Section 8.4).

Engineers will design the distribution reservoir and the pipe network to cope with peak hourly discharges. The demand for water rises rapidly at meal times, and when families finish watching a popular TV programme and all want to make tea and flush toilets at the same time. The pumping station and the treatment works function on the basis of more steady flows and are designed to handle peak daily demand in the summer.

Public health depends on clean drinking water and so water should be treated to established national and international drinking water standards. The essential requirement is freedom from bacteria which cause water-borne diseases such as typhoid, dysentery and gastroenteritis. Water need not be pure H_2O, which is rather unpalatable to drink, but it must be free of unpleasant smells, colours and chemicals likely to cause problems. Calcium and magnesium salts make the water 'hard', and soap is difficult to lather. It tends to precipitate in shower nozzles and taps, causing blockages and unsightly white deposits. Some small quantities of chemicals, however, are beneficial; for example, fluoride can be particularly good for teeth.

Treating water to make it safe and palatable includes, screening to take out large solids, settling to take out smaller suspended particles, filtration through sand beds to take out finer particles and finally disinfection with chlorine to kill the bacteria. Aeration by spraying water into the air can remove odours. Chemicals may be added to water to prevent algal growth and other chemical processes are used to remove harmful salts.

A particular problem with groundwater is dissolved iron. If groundwater contains even small quantities of dissolved iron, it comes out of solution when it reaches the atmosphere and colours the water with a bright orange slime. It looks unpleasant; the iron slime can also block pipes and filtering equipment. The iron can be removed by holding water in a reservoir for a short period while the iron settles out. The clean water can then be drained for use.

8.3.2 Unconventional water supplies

When conventional water resources are in short supply and most of the local water resources are already exploited, people start to look around for other sources of supply. These are known as *unconventional resources*, and include wastewater and desalination.

8.3.2.1 Wastewater

Treated wastewater is one potential resource. In developed countries, substantial volumes of treated wastewater are already reused and accounted for when discharged into freshwater rivers. But many countries still discharge wastewater into the sea. It is this water which is a potential new resource if it is turned inland.

Most of us do not like the idea of using treated wastewater for domestic purposes, although most people outside the water sector are unaware that this is already common practice, albeit in a very diluted form. Agriculture is an obvious alternative. But even using it to irrigate food crops can be culturally abhorrent for some. There are, however, possibilities to use it for cereals, some root crops, feedstocks, biofuels and forestry, which are less contentious.

In many developing countries, wastewater from urban dwellings is discharged untreated and is regularly used again by people who have no other sources available. Many poor people eke out a living using domestic wastewater to irrigate fruit and vegetables which they sell in urban markets. Wastewater is usually rich in nutrients and good for crops, but health risks from eating food grown in this way are extremely high and it is advisable not to eat anything that is not washed or boiled. Most indigenous people neither have the education to understand what contamination means and the impact it can have on their health, nor do they have the wherewithal to do something about it.

Using wastewater for irrigation may seem a good idea but it does create new problems. Wastewater is produced in the towns, whereas irrigated farming is often some distance from urban centres. No one wants the smell of sewage close to their homes. Canals and/or pipelines will be needed to transport the water. Also wastewater is not always available when crops need water. So storage will be needed to match supply and demand. All this can add considerably to the costs of reusing wastewater for agriculture.

8.3.2.2 Desalination

Desalination is another form of water treatment which removes salt from saline water to produce freshwater. Sea water contains 30–50 g of salt per litre and brackish water, usually a mix of sea water and freshwater in river estuaries, contains 0.5–30 g/L. Treatment methods developed over the past 30 years are well established in the water-short Near East, where it has become the main source of municipal and industrial water supply. The region accounts for over half the world's annual desalination capacity of some 6 billion m^3.

The two main processes are distillation – which heats saline water and produces water vapour which condenses into freshwater; and membrane technologies – which force saline water through semipermeable membranes to remove salt. Both processes are energy intensive and so highly suited to countries with abundant oil and solar resources. Producing good quality water can cost as much as US$ 0.5–1.5/m^3 depending on the size of the plant, salt content and energy costs. Costs continue to fall as the technologies are improved, particularly the membrane methods. Interesting though is that the freshwater produced is often too pure and not so good for drinking. Chemicals are added to make the water more 'natural' and tastier.

8.3.3 Past and future water supplies

In the UK, the development and expansion of good water supply systems coincided with the rise of civil engineering as a profession in the nineteenth century. As London began to grow and spread, the main water supply was from the River Thames and its tributaries and all were often inadequate and contaminated. Several cholera outbreaks in the city led to the Metropolis Water Act in 1852 which ordered companies to make provision for securing pure and wholesome water for the metropolis and all water was required to be filtered. In 1902, a number of small private water companies, established in the eighteenth century, where taken over and extended to form the Metropolitan Water Board.

Today, London is supplied by four private water companies, the largest being Thames Water supplying over 75% of the city's population with water from the Thames and Lea Rivers and from underground chalk aquifers west of the city. Large centralised water treatment works feed into extensive pipe networks across the city. Much of this pipework is still cast iron, installed by engineers in the nineteenth century; which is slowly deteriorating. This has led to widespread criticisms as more than 25% of London's water is lost through leakage from the network. Replacing the cast iron with modern plastic pipes is underway but it is a massive, time-consuming and costly undertaking in a crowded city.

Many other towns and cities across the world have followed a similar development pathway but in future this conventional approach may well prove too costly and unsustainable as water resources become more scarce and unreliable. The world's urban population is rapidly expanding. Since 2015 over half of the world's population now live in towns and cities. The number is expected to nearly double to 6.4 billion by 2050 and 90% of this expansion will be in the developing countries which are already experiencing major urban water problems.

Some engineers are now calling for a shift in thinking towards a more integrated approach to urban water management. A shift from highly centralised systems to more localised self-contained systems which bring together the entire urban water cycle – how water is captured locally from groundwater, streams and storm drains; how it is treated and distributed; how wastewater is collected, treated and reused; and how waste products can provide energy and nutrients for urban farming. So water supply systems seem to be going full circle from the small local units of the eighteenth century, to large centralised organisations, and now back to small locally owned water companies offering a wider range of services.

8.4 WASTEWATER TREATMENT AND DISPOSAL

8.4.1 Learning to clean it up

Freshwater and wastewater are both part of the same water resources cycle: as one person's wastewater, suitably treated, often becomes another person's water supply. It is not so long ago that household water-borne sewage, including faeces and urine, was thrown out into the city streets across Europe with the hope that open gutters and a good rain storm would wash it away, out of sight and out of mind, but usually into the local river. In the UK in 1847, civil engineer John Phillips provided the catalyst for action when he reported on the shocking conditions in London, of *'thousands of houses in the metropolis where overflowing cesspools fouled these dens of pollution and wretchedness'*. Cholera was a constant threat with three major outbreaks in the 1800s claiming over 20,000 lives. Most sewers at the time were open channels and where designed only to drain storm water.

Modern sewerage systems (sewage is domestic wastewater and sewerage is the pipe disposal system) date back to rebuilding the city of Hamburg after a disastrous fire in 1842. The destruction created an opportunity for a British engineer, William Lindley (1808–1900), to design and build one of the first underground sewers in Europe. Following this success, Lindley's skills were in demand across Europe building sewers in Dusseldorf, Budapest, Prague and Warsaw. In London, following the cholera epidemics, civil engineer Sir Robert Rawlinson (1810–1898) advocated separate sewers for rainwater and domestic sewage and laws were passed requiring all houses and cesspools to be connected to the main sewer network. And so in 1859 the foundations of an effective sewerage system for London began to take shape.

Around this time other cities were also beginning to clean up their mess. In Paris, following the ravages of a cholera outbreak in 1832, construction work began on large underground sewers which also provided conduits for other services, such as water supply pipes, and telephone lines. In Brooklyn, New York, in 1857, chief engineer Julius Adams (1812–1899) built a sewerage system which eventually became a model for other cities across the United States.

The first sewers effectively transferred the domestic sewage problem from city streets into the rivers and the sea. This dilutes the sewage and reduces the pollution; a principle that is still widely used today to varying degrees. This is okay, provided the receiving water is large enough to absorb the pollution and there is a period of time when the receiving water can recover naturally.

In the UK, most sewage is now treated before it is discharged, but there are times when storm water overwhelms treatment plants and it is discharged untreated. In 2014, at least 45 beaches were reported to be polluted by four or more sewage spills and water companies faced prosecution. In Mediterranean countries, some 80% of urban sewage is still discharged untreated into the sea because of poor and ineffective sewage treatment. In many developing countries the situation is much worse. So be careful where you go swimming.

Concerns about polluting rivers, lakes and the sea gradually shifted thinking from just piping sewage away from households to finding ways of cleaning it up. Today it is forbidden in most developed countries to discharge untreated sewage into inland waters. But just forbidding it does not always result in purifying our rivers near cities which have suffered severe pollution in the past. Adequate policing of the rules is essential and cleaning to accepted standards takes time and money.

8.4.2 In developing countries

Most towns and cities in developing countries are still at very early stages of dealing with storm water and sewage. Many do not have pipe systems and rely on open channel sewers. Conditions in poorer districts are often grossly insanitary and households are not connected to any disposal system. Sewage is discharged untreated, there is the ever-present risk of serious disease outbreaks, and diarrhoea is an endemic problem in the tropical heat. There is often little or no separation between sewers and freshwater sources: so one person's sewer becomes someone else's 'fresh' water source further downstream. The World Health Organisation estimates that more than 3.4 million people die each year as a result of water-related diseases. Most victims are young children who die of illnesses caused by organisms that thrive in water contaminated with raw sewage.

8.4.3 Sewage is...

Despite common misconceptions, usually driven by bad smell and appearance, domestic sewage is in fact 99.9% water. It is usually grey in colour and streams contaminated with raw sewage take on this colour. Today the term 'wastewater' claims to be a more comprehensive term than 'sewage', which generally refers to household wastewater. In the early days of treatment, sewage came mainly from domestic premises and its composition did not vary greatly from town to town. But life has become more complex and wastewater now comprises domestic sewage, industrial wastewaters and agricultural and urban runoff. There are no 'blue-print' approaches for dealing with wastewater. In some cases, household sewage is treated separately from industrial wastes and in others the two are mixed. Storm water is often separated because it requires less treatment before it can be discharged but it too is often mixed to dilute other wastes.

Modern chemicals now make even storm water difficult to treat. Agricultural runoff may contain fertilisers and pesticides, which are a major cause of eutrophication in lakes, and animal slurries, which are often spread onto the land as fertiliser. A particular problem is metaldehyde, used to kill slugs on farm crops, which is difficult and expensive to remove during waste treatment. The most effective ways to reduce agricultural pollution is to encourage farmers to use fewer chemicals on their land and to clean up drainage water on the farm using reed beds, prior to discharging it into rivers. So the responsibility for treatment lies with the polluter rather than with the local water authority – the principle is that the 'polluter pays'. Urban runoff also contains chemicals; oils, both natural and manmade lubricants are common. Spills during accidents with vehicles carrying a wide range of chemicals can also find their way into the treatment works and create lots of problems for the engineers.

8.4.4 Typical water treatment works

Wastewater treatment is about reducing organic content and removing nutrients and pathogens so that the resulting effluent is not a threat to public health or to the water environment.

In the UK, up to 80% of the water supplied to homes and industries eventually flows down into the sewerage system and is delivered to treatment works where it is cleaned up prior to discharging it directly into a river or the sea. Hence, the reason why water companies charge for both supplying you with water and also for taking it away. Fresh clean water is delivered under pressure in small diameter plastic pipes, usually up to 25 mm in diameter. Larger pipes, at least 100 mm in diameter, are used to take away wastewater. Sewers carry some solid material and so large diameter pipes are essential as they do not block so easily.

Water supply pipes flow full, under pressure, and at high velocity whereas most sewer pipes flow only half to two-thirds full at low velocity. They are, in effect, open channels which rely on gravity and so will only flow down-hill. If sewers flowed under pressure, there is a risk of contamination if pipes leak or burst. If gravity sewers leak, water flows in rather than out. Sometimes it is necessary to pump wastewater, and in order to avoid block-ages, centrifugal pumps have open impellers. For smaller heads and high discharges, Archimedean screws are commonly used (Figure 8.3).

The hydraulic design of treatment works involves fixing the size and layout of pipes, treatment works and associated plant based on estimates of present and future population growth, say 50 years ahead. Treatment facilities may be built to cope with the next 15–20 years with provision to expand to allow for unexpected developments of new housing and indus-tries. The system must be able to cope with peak flows which are likely to be at least twice the average daily flow.

Figure 8.4 shows the layout of a typical wastewater treatment works for domestic sewage. They are usually hidden from view as few people want

Figure 8.3 Archimedean screw pumps in Kinderdijk, The Netherlands.

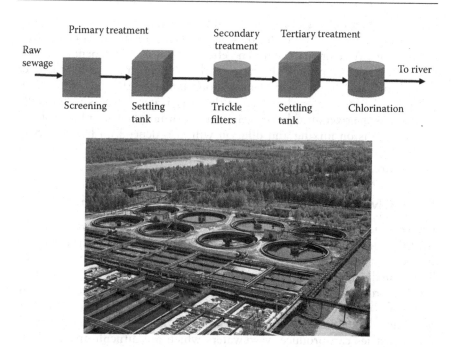

Figure 8.4 Typical layout of sewage treatment works.

to be faced daily with images and thoughts of sewage treatment. They are often most visible when taking a train journey. They are located close to rivers so that treated effluent can be easily discharged. Primary treatment begins with screening to take out large solid objects. This is followed by a settling process which separates most of the smaller solid material from the liquid. The solids can be rendered stable and almost odourless by bacterial action so that it can be dumped, incinerated or spread onto the land and used as fertiliser.

Secondary treatment is designed to reduce organic matter content and pathogens. There are several ways of doing this but a common treatment is to use biological trickle filters. These are large circular tanks with slowly rotating booms which trickle the liquid onto a deep bed of coke material. A bacterial slime builds up within the filter and any material still in suspension and pathogens stick to the film. Bacteria present in the air go to work and produces an effluent which is fairly inoffensive though by no means safe.

An alternative to trickle filters is to use *activated sludge*. Wastewater is mixed with bacteria-rich (activated) sludge and air/oxygen is pumped into the mixture. This promotes bacterial growth which decomposes organic matter and greatly reduces harmful bacteria and viruses.

Tertiary treatment uses settling tanks to remove any remaining solids and chlorination to kill any remaining bacteria. At this point, the effluent is much cleaner but is still not fit to drink and so the final step is the natural purification process which takes place when the effluent is mixed and diluted with river or sea water. The effectiveness of this depends on how clean the effluent is prior to discharge and how clean and how big the river is to absorb and dilute the effluent without damaging the river. The 'natural' process uses oxygen dissolved in the river water, but this is gradually replaced as the river absorbs oxygen from the atmosphere as it flows along. The decisions on mixing and dilution will also depend on how the river water is used downstream. It may well support recreational use, be used for agriculture, but it may also be used for public water supply.

8.4.5 Civil and chemical engineers working together

Civil engineers are mainly concerned with building the structures and solving the hydraulics problems of moving water and solids from one treatment to the next and then discharging it safely through an outfall structure. Traditionally, wastewater treatment works were designed on an 'end-of-pipe' approach, which means that everything went into the treatment works and what came out was fit to discharge into the river or sea. This was fine when effluent was mostly 'standard' domestic sewage. But modern industries can produce wastewaters which are difficult and costly to clean up. So the onus is now placed on the industries to clean up their wastewater to approved standards before discharging into the public sewers. But even domestic wastewaters have changed as new products have become available, like synthetic detergents, which have modified the traditional approach. It is also surprising to see the large amount of oils and fats poured down domestic sewers even though it is a serious offence to do so. Simple oil traps would eliminate this problem.

Wastewaters are now so complex that chemical engineers are as much involved in planning and designing water treatment processes as civil engineers. The approach is no longer big concrete settling basins but a series of closely integrated units which provide both physical and biological treatment tailored to deal with individual wastewaters.

8.5 IRRIGATION ENGINEERING

8.5.1 Irrigation's beginnings

Although much of our food is grown using natural rainfall, irrigation has enabled many of the drier parts of the world to become highly productive. Irrigation is not new. It has been practised for many thousands of years and was the mainstay of early societies, such as in Egypt, China, India, Iraq and Syria (Mesopotamia): often call 'hydraulic societies' because of their reliance on canal systems taking water from some of the world's great rivers. Historical records show that these where sophisticated and well-managed systems with rules and regulations enforced by law. The most well-known is the Code of Hammurabi from Babylon, which set out irrigation water allocations and the penalties for stealing water from fellow farmers. The Babylonian civilisation relied entirely on irrigated farming as there was virtually no rainfall in that region. But the society lacked understanding of salinity which plagues irrigation in arid regions: how harmful salts, dissolved in the river water, can build up in the soil over a period of time and kill crops. Ancient records show a gradual change in cropping from wheat to barley, which is a classic sign of salinisation as barley is more salt tolerant than wheat. But eventually the whole civilisation died out when even barley would not grow.

The ancient art of irrigation also spread westwards into the drier regions of Greece, Italy, Spain and North Africa. As the big hydraulic societies faded, Spain became one of the reservoirs of irrigation knowledge for many hundreds of years. Modern irrigation practices began to emerge in the late nineteenth century when British civil engineers, like Sir Claude Inglis (1883–1974) and William Wilcocks (1852–1932), who built the first Aswan dam in Egypt, began developing large irrigation systems across, what was then, the British Empire to benefit millions of people living in abject poverty. Engineers travelled to Spain to learn about irrigation and transferred the knowledge and experience back to Egypt, Mesopotamia and India.

8.5.2 Irrigation today

Irrigation today is one of the success stories of the twentieth century. It makes a major contribution to feeding the world's rapidly growing population which doubled last century to 6 billion. Some 1,500 million hectares of land are cultivated to provide the global community with food and natural fibres, such as cotton. Only 300 million hectares (20%) are irrigated yet this area produces more than 40% of the world's agricultural output. Irrigated crops can produce up to 3 or 4 times the yield of rainfed crops, particularly in the drier regions where rainfall is seasonal, meagre and unpredictable. In the 1970s the 'Green Revolution', as it was known in Asia, transformed rice production by developing new varieties, using fertilisers and controlling irrigation water. Rice production increased from 2–3 tons per hectare (ton/ha) to as much as 6–8 ton/ha and more in some countries. A great many people had more food to eat and improved livelihoods but there was criticism over the potential impact all this had on the aquatic environment.

Much of South Asia continues to rely on water from the Indus and Ganges rivers. Irrigation in India has greatly improved food security, reduced the dependency on uncertain monsoon rains, increased crop yields and created many thousands of rural jobs. More than 58 million hectares are irrigated which is over one-third of the cultivated land area.

Egypt still relies entirely on the River Nile to irrigate crops; Mesopotamia, now Iraq and Syria, relies on the Tigris and Euphrates Rivers; and China's

---

BOX 8.3 IRRIGATION IN THE UK

In the UK, rainfall is sufficient and timely to grow most basic food crops. But in the east and south east of England, where the soils are sandy and the climate is ideal for growing fruit and vegetables, rainfall is low and unpredictable and so farmers irrigate to make sure they produce good yields and high-quality crops on time for the supermarkets. This is now a highly specialised and profitable market. Indeed, the UK has been described as one of the most sophisticated food markets in the world. However, water shortages and the fear of droughts are raising concerns among farmers who are now building storage reservoirs to conserve water when it is plentiful for times of shortage.

---

main source of water for growing food in the dry north is the Yellow River. Irrigation is also important in Australia, Indonesia and western United States.

Even in humid and temperate areas, such as Europe and central Africa, irrigation makes an important contribution to food production. Crops can be grown under natural rainfall but often this is unreliable and poorly distributed and droughts are becoming more prevalent. Under these conditions irrigation is used to supplement rainfall to improve crop yields and quality (see Box 8.3).

Globally, irrigation has emerged from an art, practised in ancient times, into a science. Engineers have learned to calculate the water requirements of crops (see Box 8.4), assess water availability, design major engineering works to store and control some of the world's great rivers and develop elaborate systems of canals and structures to control and transfer water from the rivers right down to individual fields.

8.5.3 Crops needs lots of water

Food crops need lots of water. Indeed, some 70% of all water abstracted globally from rivers and aquifers is used for agriculture. This is a surprising figure compared with the rather modest demand of less than 10% for domestic water supply. Most people have little idea of the vast amount of water that goes into producing our food. In an arid country like Egypt, it takes about 1,200 mm depth of water applied gradually over the 3-month growing season to produce a crop of maize or potatoes. This is 12,000 m³/ha: enough to provide the daily domestic water supply for 80,000 people in Europe.

8.5.4 Irrigation schemes

Irrigation schemes come in all shapes and sizes. In developing countries, many millions of smallholder farmers cultivate small plots of land (1–2 ha)

BOX 8.4   PENMAN EQUATION: HOW NECESSITY IS THE 'MOTHER OF INVENTION'

The British meteorologist Howard Penman (1909–1984) worked out how to calculate the amount of water a crop needs to grow and produce a good yield. Although this is the foundation for designing irrigation systems, it was not the reason for his research. In 1944, the British government recruited Penman to work out the physics of how grasslands consume water through a process of evapotranspiration in order to prepare for the 1945 allied land campaign during World War II. Crucial to the success of the landings was to work out how long after rainfall would the heavy clay soils in Europe be strong enough to take gliders and tanks without them sinking into the mud? Soils get stronger when they dry out and so he worked out how much water the grasses were transpiring from the soil each day. From this he was able to calculate how many days it would take after rainfall for soils to dry out and become strong enough to take the loads.

Today, the Penman equation is the internationally accepted method for calculating crop water requirements for irrigation. It is the basis on which most irrigation systems are designed. It is also widely used for assessing water use of all kinds of vegetation as part of water resources management. It is an excellent example of how classical physics can be used to solve real environmental problems.

---

and privately invest in irrigation technologies, in particular small pumps for exploiting streams and shallow groundwater. But there are limits to how many farmers can occupy the river bank and pump water and so alongside the smallholders are the big formal irrigation schemes which cover many thousands of hectares: similar in many ways to those developed by the ancient hydraulic societies.

In the 1970s and 1980s, international aid agencies invested heavily in large-scale irrigation in developing countries as a means of alleviating poverty and increasing food security (Figure 8.5). This was technology led – engineers designed and built schemes to supply water to farmers who would then get on with the job of growing crops.

Typical irrigation schemes include dams to store water, major river control structures, headworks to divert water for irrigation, pumping stations to lift water to irrigate higher ground, large open channels, hydraulic structures to measure and control water distribution and dissipate unwanted energy, culverts and bridges for access, and drainage systems to take away excess water and avoid water logging and salinisation.

Most large-scale irrigation systems use open channels (canals) rather than pipes. Canals are cheap, easy to construct and well suited to conveying

(a)   (b)

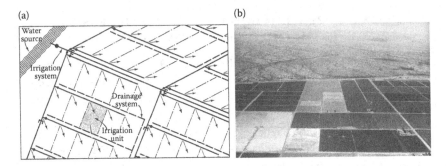

Figure 8.5 Typical large-scale irrigation scheme (a) Diagram of irrigation and drainage layout. (b) Aerial view of a large scheme.

large discharges on gently sloping land where most schemes are built (see Chapter 5). Some canals are lined to reduce seepage losses and maintenance. Canals also provide access to villages and farms. So as well as their hydraulic design, other key features include making sure the embankments are wide and strong enough for vehicles to use for access and maintenance.

Hydraulics structures, both large and small, control discharges and water levels throughout a scheme from the river to the farm (see Chapter 6). Water levels, known as command levels, are important to maintain gravity flow from the canal system into the fields and avoid extra costs of pumping. Different countries have adopted different approaches to canal control, influenced mainly during colonial times. Schemes built by British engineers, for example, tend to use movable gates which allow flows to be adjusted to meet farmer demands (Figure 8.6). In countries influenced by Dutch engineers there was a preference for fixed weir structures, which are less flexible but much simpler to operate. French engineers, influenced designs in North Africa where water is scarce. Their approach was to use automatic control gates which adjust water levels and discharges as demand changes. These are usually more expensive but can be less wasteful of water.

(a)   (b)

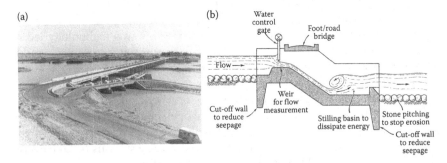

Figure 8.6 Major barrage structure in Iraq diverting water into irrigation main canal. (a) Barrage and canal headworks. (b) Typical cross-section of barrage structure.

8.5.5 Methods of irrigation

8.5.5.1 Surface irrigation

The most common method that farmers use for spreading water across their land is surface irrigation. This method is used on about 95% of all the globally irrigated land area (Figure 8.7). Most farmers use basins which are large flat areas and ideal for rice growing. Some use borders and furrows, which work well on gently sloping land for a wide range of row crops.

Surface irrigation is usually the cheapest method and is thought to be the simplest to manage, but it has a reputation for wasting water. This reputation may be deserved in developing countries where irrigation is in the hands of poorly educated peasant farmers working on small plots and who may have no concept of what the word 'discharge' means. But it is undeserved in countries, like the USA, where 40% of irrigation is still with surface methods and farmers have large fields, more control over flows and managers are aided by computer-based flow models.

8.5.5.2 Sprinkler irrigation

Sprinkler irrigation accounts for most of the remaining 5% of the global irrigated land area. It relies on pressurised pipes carrying rotating spray nozzles to spread the water (Figure 8.8a). The first commercial systems were developed in the eastern United States in the 1930s as a means of moving irrigation systems from one field to another to supplement rainfall. This is something that you cannot do with surface irrigation. It came about as aluminium pipes, which can be easily moved around fields, were beginning to be manufactured; and centrifugal pumps, which provide the ideal discharges and pressures for sprinklers, were being mass produced. This simple system of moveable pipes and pumps is still very popular today, but mostly on small farms where farm labour is not a problem. But on larger farms and the difficulties of recruiting labour, much of the technology developments over the past 80 years have been to automate sprinkler

(a)   (b)

Figure 8.7 Surface irrigation. (a) Basin irrigation. (b) Furrow irrigation.

(a)   (b)

Figure 8.8 Sprinkler irrigation. (a) Small rotary sprinklers. (b) Centre pivot irrigation machine.

systems. The ultimate machine is the centre pivot which can irrigate up to 100 ha and one person can operate several machines. These produce the famous green rings which can be seen when flying over desert areas. Irrigators claim that this machine has transformed irrigation in a way that tractors transformed agriculture (Figure 8.8b).

According to farmers 'pressure is king' and so the key to successful irrigation is to operate at the recommended pressure and discharge. So irrigation engineers select pumps, pipes and spray nozzles that meet the pressure requirements. In this way, it is possible for sprinklers to provide good uniform watering and for spraying water to have a wide range of drop sizes which do not damage soils and crops.

8.5.5.3 Trickle irrigation

Although trickle irrigation accounts for less than 1% of world irrigation, it is an important method for some countries, but is not as commonly used as people think. The method relies on small, frequent (sometimes daily) water applications from emitters which drip water from a pressurised pipe system (about 1 bar) so that only the soil around each plant is wetted (Figure 8.9). Blocked emitters are a big problem and so water must be filtered to take out sand and silt particles, and treated chemically to stop calcium and iron from precipitating in the emitters. A big advantage is being able to apply fertiliser mixed with the irrigation water and applied directly to the plants. This may be the reason why trickle manufacturers sometimes claim bigger yields with trickle irrigation.

8.5.5.4 Which method?

There are various physical factors which constrain the choice of irrigation method, such as land topography, soils, crops and labour availability. Surface irrigation is ideal on flat or gently sloping flood plains, and particularly for

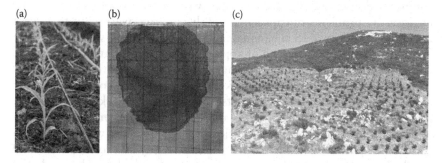

(a)   (b)   (c)

Figure 8.9 Trickle irrigation. (a) On a row crop. (b) Soil wetting pattern. (c) Trickle on avocados.

growing rice which is one of the world's staple crops. Sprinklers are useful on undulating land and sandy soils but can be a problem when it is windy. Trickle is ideal when both land and water are poor in quality and in short supply, and crops have high market value. One avocado farmer used trickle irrigation and turned waste land on his farm into a highly productive enterprise (Figure 8.9c).

Surface irrigation is seen as the least efficient method; followed by sprinklers, and trickle as the most efficient. But this is misleading as efficiency depends as much on the management ability of the farmer as on the technology *per se*. Trickle irrigation is much admired because of the claims made that 'trickle uses less water'. This is a rather vague statement and implies that somehow crops actually use less with this method, which is not the case. Crops respond to water and not to the method of applying the water. Savings are possible but these come from reducing water wastage which is more likely with surface and sprinkler irrigation when water can seep deep below the roots or run off the land and into the drainage system. Trickle is the costliest of the three methods and ultimately there is the problem of disposing of large amounts of plastic pipe at the end of the irrigation season.

Trickle irrigation also presents a psychological problem for farmers. In the United States, farmers say they 'like to see the water'; hence the preference for surface and sprinkler methods. But with trickle all you get is a small wet patch on the soil surface and a hope that the soil is wetting up properly underneath. The farmer's nightmare is the partially blocked emitters that show a wet patch, which looks fine but the plants may not be getting enough water. The farmer only finds this out when his yields are down. So using trickle irrigation requires a lot of careful management.

8.5.6 Managing irrigation systems

Like many areas of water resources engineering, large-scale government-run irrigation schemes up to the 1970s was all about technology. Irrigation engineers dominated the sector. They were good at solving the major engineering problems of capturing and distributing water. But they were criticised for taking over the management of systems, for which they were not trained, and for their lack of appreciation of the end use of irrigation. They knew little about soils and agronomy, mechanisation, livelihoods of farming families and local markets where produce could be sold for profit. Agronomists and economists where meant to take care of these issues but equally they had little appreciation of the engineering constraints implicit in operating irrigation systems. In other words, the main cadres of engineering and agriculture did not really talk to each other and saw little need to do so. This is now changing but it is slow. The result was that many schemes were poorly funded, managed by inexperienced staff, poorly maintained and criticised for being inefficient in terms of water use and producing crops.

Indeed, some schemes were running below 50% overall water use efficiency – the ratio of water actually consumed by the crop in relation to the amount diverted into the scheme. So at least half the water withdrawn from the source appeared to be lost. Strangely, few people seemed to question where the 'lost' water was going and whether the loss was in fact benefiting other users downstream (see Section 9.6.3).

All this criticism led international aid donors and government investment to move away from irrigation. So smallholder farmers began to take matters into their own hands rather than waiting for governments to do something. In India, many farmers abandoned the canal systems which were unreliable and drilled shallow wells and began pumping groundwater. In fact, the leaking canal systems were recharging the aquifers that farmers were pumping from. Today over 60% of all irrigation in India is from groundwater, mostly private individuals supplying commercial markets with food products. So successful has this been that India now has a problem of regulating groundwater abstraction rather than a surface water problem.

Irrigation has been less successful in sub-Saharan Africa. Though in some parts smallholders successfully irrigate their farms and use local surface and groundwater to irrigate crops for local markets. Some use small petrol-driven pumps and others use treadle pumps (see Section 7.2). The reality is that many millions of poor people continue to survive on subsistence farming, yet the continent is well-endowed with land and water resources waiting to be exploited. Some commercial sugar, tea and coffee estates are successful but smallholder government-run schemes often fail because they are too costly to build, and lack trained support staff and resources. Food marketing systems are poor across the continent and farmers do not have efficient outlets for their produce. So perhaps it is not surprising that farmers, unlike their Indian and European counterparts, grow mainly for their own family needs rather than for profit. Staple food crop yields are stuck around 1 ton/ha and have been so for the past 60 years in spite of massive investments in soil, crop and water research to increase productivity.

8.5.7 Irrigation: the downside

Irrigation does have its downside and it is often blamed for large scale environmental damage. Large areas of India, Pakistan and Bangladesh have serious salinity and water logging problems which stem from poor irrigation and drainage practices. But what is often ignored is that many millions of people in the region are alive today because of the prosperity and food security that came with irrigation to communities that may otherwise have died from starvation. In Central Asia, the environmental damage to Aral Sea is often laid at the door of irrigation engineers. There is little doubt that they are part of the problem but more so are many others including politicians in the former Soviet Union who wished to develop those large areas of irrigated cotton to the detriment of the local environment.

Besides the risks of salinity and soil erosion there are health risks to consider, particularly as malaria, carried by Anopheles mosquitoes, and schistosomiasis (bilharzia) carried by snails, thrive in the tropics wherever there is slow moving water. Both kill many thousands of people each year and are considered to be the 'diseases of irrigation'. Engineers can do much to control the diseases by manipulating the environment in canals and fields so that they are hostile to the 'vectors' (mosquitoes and snails) which carry the diseases.

Clearly, irrigation still has a major role to play in feeding the world but developing schemes is complex and fraught with difficulties. There are two traps to fall into. The single criterion trap says that we continue to build schemes and focus mainly on the engineering or the agronomy. This means people can get on with the job in hand and not worry too much about its relevance to the bigger picture. The other is the all-inclusive trap which says that we must bring in all the issues if a scheme is to be relevant and successful. But this only leads to 'paralysis by analysis'. Engineers, with their culture of making choices and negotiating trade-offs, together with the many other disciplines involved in irrigation development, have a great future here in choosing the most appropriate pathway between these two extremes.

8.6 RIVER ENGINEERING

Believe me, my young friend, there is nothing – absolutely nothing – half so much worth doing as simply messing about in boats.

Wind in the Willows

This is perhaps the most quoted comment about rivers in English literature. Rivers offer many benefits. They provide water for municipalities, irrigation, drainage, navigation and amenity but they are also attractive and environmentally beneficial in the landscape, and great for messing about in boats.

Left to themselves rivers choose their natural meandering course as they flow through a catchment to the estuary and the sea. In the UK in the mid-twentieth century, the fashion was to 'engineer' rivers to be efficient channels which improved land drainage and reduced urban flooding. Channels were straightened and in some cases lined to increase flow velocities.

Modern river engineering is very different. The natural bends are being put back and the concrete linings removed to recreate natural channels as we have come to realise that rivers are not just convenient water carriers but provide habitats for all kinds of flora and fauna which enrich our lives and also our water resources. This new approach is known as *environmentally sound river engineering* (Figure 8.10). The governing principle is to treat the river in a holistic way before taking action to solve a local problem. This approach takes into account not just the hydrology, hydraulics, sediment transport, river services and maintenance, but also river morphology, wildlife, landscape, amenity, angling and aquatic biology. Legislative issues and planning requirements are also important – who has permission to use the river and/or abstract water and for what purpose? Who can discharge water into the river and what standards of water quality are acceptable? Planners now accept that rivers are not just water conduits for drainage and that developing plans for rivers, or catchment management plans, will involve all the interested parties – the stakeholders. This has marked the shift from regulating the rivers for people to regulating the people for the benefit of the rivers.

Compared to the hydraulic design of pipes and channels, river engineering is not so precise as there are still gaps in our understanding of the links between discharge and the characteristics of natural rivers. In the absence of adequate theories, engineers supplement their theoretical knowledge with judgements based on extensive practical experience. But there are many pitfalls for the river engineer who is often accused of 'interfering with nature'. When bridge abutments encroach into a river channel they can cause serious bed and bank erosion, cutting off a loop to straighten a river can cause scouring as the channel slope increases plus siltation

(a)   (b)

Figure 8.10 Natural river (a) vs. an 'engineered' river channel (b).

downstream, and a scheme designed to avoid flooding at one point may well create flooding elsewhere.

What is known is that the cross-section of rivers flowing naturally in alluvial valleys tends to be wide and shallow, particularly on flood plains, and develop a stable shape – bed width, depth, slope – which relates to their dominant discharge, the characteristics of the surrounding soil and the silt being carried by the flow. Gerald Lacey (1887–1980) and other civil engineers, involved in building the major irrigation schemes in India during the colonial period, developed a set of empirical equations based on observations of stable irrigation canals. These link natural channel characteristics to discharge and provide a rational approach to designing irrigation canals which carry a light silt load. They became known as the 'regime equations' and were found to be useful also for designing engineering works in natural river channels in much the same way that the Manning and Chezy equations are used for man-made channels carrying clean water.

Regime equations have also proved useful to bridge designers as they provide a good estimate of the 'natural' width of a river. Engineers often try to keep costs down by shortening a bridge crossing and constricting the river channel. But it is not unknown for engineers to build bridges across flood plains only to find that the river has changed course and left the bridge high and dry. It seems that the essence of good river engineering is to work with the river rather than against it.

River engineering works are used to regulate flows and river levels for water supply, hydropower, flood control and navigation. In tidal estuaries, barriers can limit salt water contamination upstream. Building dams can bring a number of benefits, although in recent years there has been much concern over the land that is flooded and the people that are displaced from the reservoir area. Dams provide storage to even out fluctuations in river flows and regulate supplies downstream. If a dam is high enough it can provide enough head to run turbines to generate electricity or a gravity supply in a pipeline or an irrigation canal. If the dam storage is not full then it can store storm water and take the peak off flood flows which would otherwise cause damage downstream.

Navigation is also an important consideration on some rivers. Although rail and road haulage now dominate transport services, many rivers, such as the Rheine and the Danube, are used for water transport. Indeed, it is possible to travel by boat from the Rheine's estuary in the North Sea through to the Danube's estuary in the Black Sea. Locks help to maintain water levels. Much of the grain harvest in central Europe is transported in barges down the Danube to the Black Sea for export. Droughts impact river traffic and in 2012 there were concerns over maintaining water levels along the river to keep the grain barges moving.

Natural channels do have a tendency to meander in a sinusoidal path in alluvial valleys which means that continuous maintenance is needed where the river cuts away at the outside of bends and deposits silt on the inside (see Section 5.9). This is made more complex in tidal rivers when flows are regularly reversed. Maintenance includes building revetments from stone or timber to protect banks from erosion and periodic dredging to maintain the main channels, particularly for river traffic.

8.6.1 Modelling rivers

Because of the complexity of flow conditions in rivers, major works, such as flood alleviation are often model tested prior to taking action. In the past, physical models were built in laboratories to enable lots of possible design options to be tested to see which provided the most appropriate solution. By modelling the flood plain it was possible to see and measure the extent of flooding and assess the impact of new engineering works both locally and downstream. For many years the laboratories at HR Wallingford, UK had a physical model of Morecambe bay in Lancashire with various rivers feeding into the estuary. It was quite a privilege to take a 'stroll' around the bay in just a few minutes and see the impacts of various river flows and tides on proposed engineering works to alleviate flooding.

Physical models are still constructed but they are expensive and many now opt for mathematical, computer-based modelling. Unlike the physical models, these work at full scale and so do not suffer from the scale effects. Surface tension forces, for example, affect the performance of small-scale physical models, whereas they are not relevant in the real situation. Mathematical models can be less costly and many alternatives can be tested quickly once the model is built. But they are only useful when the physics of flow are known and can be built into the model. The lack of topographic information can limit accuracy. From an engineer's perspective they are not quite as exciting as a physical model and convincing clients to pay for and undertake work can be easier when they can physically see what is happening rather than peering at lots of sophisticated graphics!

8.6.2 Protecting the aquatic environment

Finally, there are many planning rules and legal rights to consider which are designed to protect our water resources. In Europe, not least of these is the Water Framework Directive (WFD). This is a legally binding agreement for all countries in the European Union (EU) to collectively tackle the growing problems of deteriorating water quality, loss of aquatic ecosystem function in our rivers and water bodies and increasing water scarcity across Europe. All this stems from a growing realisation that our rivers and wetlands provide the freshwater we all need and if we continue to pollute them we reduce their natural capacity to continue producing freshwater for us in the future. On many rivers we are in danger of getting into a vicious circle of over-abstraction to meet increasing demand, increasing pollution from inadequate wastewater treatment, which further reduces freshwater availability, and drives over-abstraction. Since WFD was approved in 2009, all rivers are expected to reach 'good' ecological and quality status by 2015 or at least countries must justify why and what steps they are taking to remedy any problems. None of the EU countries have so far reached 'good' status in all their rivers which shows what a major challenge this is even in the developed, richer countries. Engineers are at the heart of this process of river improvement which concerns cleaning up discharges into rivers from domestic sewage treatment works, industries, storm water and agriculture; and managing demand in order to sustain our aquatic ecosystems.

The problems of river management and pollution are usually more acute in the developing countries where rivers are often seasonal depending on the rains. Some 20% of the world's rivers run dry before they reach the sea. Pollution in rivers from untreated sewage is serious not just because of the immediate impact on public health but also the long-term damage to aquatic ecosystems. Engineering plays a key role in meeting the challenge of providing freshwater in such circumstances but the solutions go well beyond the bounds of engineering.

Chapter 9

Water resources planning and management

9.1 THE WATER CHALLENGE

Water is an integral part of everyone's daily lives, yet each of us values it in very different ways. In the UK and other rich countries, we very much take water for granted. It is well managed, and planning and investment have greatly reduced the risks of water shortages and damage from flooding. We turn on the tap and water flows. We flush toilets and away it goes to be cleaned up. We enjoy the aquatic environment of rivers, lakes and wetlands, which add to the beauty of the landscape and provide us with leisure and pleasure activities.

Most people have little idea of the vast amount of water we use to maintain our life style. But this is beginning to change. We have seen how flooding causes misery when homes and businesses are badly damaged. In the UK, in 2011 and 2012, we saw how drought threatened water supplies to our homes and can limit the water available for basic hygiene. It raised public awareness about keeping our rivers clean, and concerns grew over agricultural and industrial wastes which, if left untreated, can pollute our water resources.

In the developing countries, people are much more aware of water in their daily lives. Water resources are often scarce. They are not well developed and severe shortages are commonplace, as much from the inadequacies of the water supply systems and poor management as from a lack of water. Wastewater from households and industry flow untreated into rivers, polluting freshwater (Figure 9.1). Floods and droughts are a normal way of life for many millions of people and thousands have died from their impacts. In some countries, people are so used to being flooded that they have developed housing which they can raise or lower depending on the river levels (Figure 9.2). In many African cities, water supplies are so erratic that wealthier people buy pumps to fill elevated tanks when water is available. Others pump it out of the main supply pipelines when the pressure is too low. Many poor people do not have access to piped water and have to buy it by the bucket from water sellers. They pay a high price for the privilege. In rural areas, people, mostly women and children, walk several kilometres each day to fetch water for their basic family needs.

Figure 9.1 Sewage being discharged into a watercourse in West Africa.

Figure 9.2 Floating houses enable people to cope with frequent flooding.

In 2015, the World Economic Forum put water at the top of the world agenda (see Further Reading). It highlighted water as the greatest risk facing our world – greater than the threats from terrorism and migration. But why has water become such a critical issue? The threats come from many sources. First is the realisation that water resources are limited and yet the demand for more water is increasing everywhere. There are concerns about rapidly growing and urbanising populations. People are changing to more 'water-rich' diets and lifestyles. There are increasing and competing demands for water from agriculture, industry and energy. Climate change brings unpredictable risks to water resources and concerns about environmental degradation. And tensions grow over scarce water resources that flow across national boundaries when each country wants a share.

Indeed, if the world continues to use water at current rate, it is estimated that demand could outstrip supply by as much as 40% by 2030. This would put both water and food security at risk, it would constrain sustainable economic development and degrade the aquatic environment on which everything else depends.

So what are we doing about it? Clearly, the 'business-as-usual' approach is not an option and changes are needed in the way we plan resource use for the future. The first step is for everyone to recognise the threat. Those working in the world of water have known about this for many years. But the breakthrough, that puts water security at the top of the list, came in 2015 when all 193 member countries of the United Nations agreed that water was at the heart of its new global development agenda to achieve a better, more just, peaceful and sustainable future for everyone by 2030. The United Nations set out 17 Sustainable Development Goals (SDGs) covering all aspects of people's lives and water is mentioned in 16 of them. But it is specifically referred to in SDG 6, now known as the *Water Goal,* which says that we must *'ensure availability and sustainable management of water and sanitation for all'* (see Further Reading).

The UN agenda recognises that water is now critical for sustainably managing natural resources. It is embedded in all aspects of development; in food security, health and poverty reduction and in sustaining economic growth in agriculture, industry and energy generation. This is often called the *green growth* or the *green economy.* The key word is sustainability, which means we must not overexploit the resources on which we all depend. Engineers often refer to the aquatic environment, which provides us with water, as the *green infrastructure,* to stress its importance alongside the more familiar water engineering infrastructure which enables us to monitor, manage and control water resources.

This is the big picture of water resources management and this chapter attempts to set out some of the basic principles and practices of water resources planning, in particular some of the myths that plague current thinking about water management. The story begins with some basic hydrology.

9.2 SOME BASIC HYDROLOGY

Hydraulics is the science of how we control water, how we store it and move it around, whereas hydrology is concerned with water in its natural state, its origins and movement over and through the Earth. In the past, hydrology used to be covered in a few chapters added to books on hydraulics. But it has now grown into a sophisticated science in its own right which measures and models each and every element of water's natural journey. The boundary between the two disciplines, however, is blurred as hydrology now embraces the physics of runoff, percolation, evaporation, river hydraulics, groundwater flow and water quality.

Hydrologists gather and process valuable data to assess how much groundwater and surface water is likely to be available now and in the future so that engineers and planners can make decisions about future infrastructure.

9.2.1 The hydrological cycle

The science of hydrology is all about the elements that make up the *hydro-logical cycle* (Figure 9.3). The cycle begins when moisture evaporates from oceans and lakes, and winds carry it over both land and sea where it precipitates and falls as rain or snow. Most of it falls into the sea. On the land, some 70% of the rainfall goes back into the atmosphere as it evaporates from rivers and lakes and transpires from forests, pastures and crops. The rest finds its way overland into rivers or infiltrates underground into aquifers and eventually ends up in the oceans again. Some water returns to the sea in a matter of days but some, particularly groundwater, can take years and even centuries to reach the sea. All these elements can now be quantified and together they provide a *water balance*, which is similar to the continuity equation in hydraulics (see Section 3.4)

Inflow(rainfall) = outflow (from evaporation, transpiration, rivers,
groundwater) + rate of increase/decrease in storage
(in dams, soil water, wetlands, groundwater).

Only a very small amount of the world's water (3%) enters the *hydrological cycle* and less than 1% (of the 3%) accounts for freshwater lakes, rivers and water in the top 1 m of soil which is exploited by vegetation (Box 9.1).

Like hydraulics, hydrology has only really developed as a science in the last few centuries. In biblical times, the link between rainfall and river flows was not well understood. Although ancient civilisations, like those in Egypt and Mesopotamia, where adept at exploiting water resources;

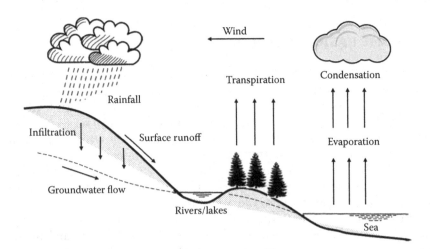

Figure 9.3 Hydrological cycle.

BOX 9.1   SOME DATA ON WATER RESOURCES

97% of the world's water is seawater.

3% is freshwater evaporated from the oceans which enters the hydrological cycle.

Of this 3%:

75% is tied up in the polar ice caps and glaciers.

Most of the remaining 25% is groundwater.

Less than 1% makes up freshwater lakes, rivers and water in the top 1 m of soil.

---

such as building water supply systems, dams, and irrigation networks; they could not believe that the meagre rainfall they experienced accounted for the flow in Nile and Euphrates Rivers. Records show that people were well aware that rainfed cropping was a risky business, whereas irrigated cropping reduced uncertainty and enabled farms to flourish on otherwise unproductive land. But the 'world' was a much smaller place in those days and so people were not thinking about catchments in areas with much higher rainfall many hundreds, even thousands, of kilometres away. Theories abound in early scriptures about rivers emerging from under the ground, which is not far from the truth as we know that underground 'springs' feed some rivers with enough water to keep them flowing all-year round. Even when modern science emerged in the seventeenth century, scientists were still puzzled as to why rivers continued to flow long after the rainfall stopped. Pierre Perrault (1608–1680) finally resolved the matter by measuring rainfall and river flows in the Seine basin in France. He measured the volume of rainfall across the river catchment and concluded that this was far greater than the volume flowing in the river and more than enough to keep it flowing long after the rainfall stopped. From this emerged the idea of the *hydrological cycle* that we know today which describes the continuous movement of water on, above and below the surface of the earth (Figure 9.3).

9.2.2 Water basins: the ideal planning unit

Once the concepts of the hydrological cycle and the water balance are understood, it makes sense to plan and manage water in terms of *water basins*. Sometimes they are called *river basins, drainage basins* or *catchments*. A basin is a self-contained, bowl-shaped area, which captures rainfall, and the rivers in the valleys provide the natural drainage channels taking water back towards the sea (Figure 9.4). The *watershed* marks the boundary around the edge of the basin, and within this unit, we can assess the water balance and determine the amount of water that we can take out and usefully use.

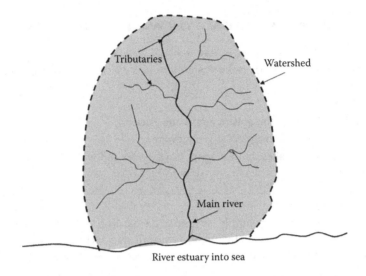

Figure 9.4 Water basin.

The basin approach is ideal for planning surface water resources, but not so good for managing groundwater. An aquifer may lie under several basins. It may be recharged by one basin and water abstracted in another. So those exploiting groundwater may need to cooperate with others to ensure that a shared aquifer is not overexploited.

9.2.2.1 Open and closed basins

River basins are sometimes referred to as *open* or *closed*. When there is sufficient water to satisfy demand even in the dry season and water continues to flow out into the sea, the basin is described as *open*. But as growth continues and water supplies dwindle, eventually there will be no usable water flowing out to sea. At this point, the basin becomes *closed*. When basins approach closure, there are often what we call *top-ender-tail-ender* problems, as those living at the tail end of the system tend to get less water of poorer quality. This can cause lots of problems for water managers. Over 20% of the world's population lives in urban communities along the coast and large rural populations rely on agricultural lands in the river deltas, all demanding more and cleaner water supplies.

New demands for water in a closed basin can only be met if someone in the basin is willing to give up some of their supply. Other options to reopen the basin include increasing storage and transferring water in from adjacent basins.

9.2.2.2 Approaches to basin management

Basin organisations are usually set up to manage water within basins but it is not always the case. Water is often managed within administrative boundaries, such as local authorities, rather than hydrological ones. This can make sense to local administrators who want to exercise control over resources within their district, but it makes no sense from a water resources point of view. You can imagine the endless confusion and conflicts when two different authorities start independently planning to use the same water, which often happens.

This problem is even greater when water basins are spread across two or more countries. Managing water which crosses international boundaries is fraught with difficulties, especially when the countries are not on friendly terms and have different legal systems and cultural attitudes to water. Currently, there are more than 260 internationally shared watercourses involving over 70% of the world's population. In Africa, 90% of surface water resources are shared. In Europe, rivers criss-cross many countries and the Danube basin offers a good example of international cooperation and water sharing. However, there are very few basins in which countries successfully share water and conflict is often the norm. International water law is helping to resolve this problem but it is one that is not going to easily disappear.

9.2.3 Selecting 'design years'

Engineers are particularly interested in the relationship between rainfall across the basin and the amount of water that runs off the land and drains into rivers. This provides information for assessing water available for domestic supply, power and irrigation, and for designing reservoirs, dams, spillways and flood alleviation schemes. Drought periods, which result in low river flows, will determine the volume of water that can be relied upon, and its availability over the year will determine the amount of reservoir storage needed. Prolonged and intensive rainfall, which results in high river flows, will establish the risks of flooding cities and farmlands, and will determine the size of hydraulic structures, such as spillways, control and diversion structures, culverts and river crossings.

Engineers always want to know what happens in extreme conditions so they can be confident that water infrastructure will perform well most of the time. So hydrologists look back at what has happened in the past and use the data to predict what may happen in the future. So good reliable meteorological and river flow data collected over long time periods are essential for predicting the future with some degree of certainty.

Hydrologists analyse rainfall and runoff data to find extreme events, such as intense rainstorms, floods and droughts. These events are expressed in terms of their probability of occurring – how likely is it that a particular event may happen. Will an event reoccur 1 in 10 years, 1 in 100 years or in extreme cases, 1 in 500 years or more. These probabilities provide the *design years* on which engineers can base their designs. They are not actual recorded years. They are 'constructed' years based on the recorded events which represent severe conditions. The idea is similar to *design loads* that structural engineers use to design buildings. They do not concern themselves with where you plan to put your bed, machines, desks and bookcases, rather they use standard loads for designing homes, factories and offices which cover all eventualities – within reason of course. The floor in your bedroom will be designed for a standard domestic load which includes a bed. But if you decide to turn your bedroom into a library or a workshop then beware, the floor may need strengthening!

Engineers and hydrologists fully understand the idea of *design years* but they are not well understood by the general public. Designing a flood alleviation scheme for a 1 in 100-year flood event does not mean that once a flood of this magnitude occurs it will not happen again for 100 years. Indeed, it is not uncommon for a similar flood, or one much worse, to occur the following year, much to the consternation of those directly involved. It is simply a measure of severity of flooding agreed upon for design.

Costs and benefits also play a role in choosing the *design year* (see Box 9.2). Money is usually limited and some options, which give high levels of protection, may well be too costly for something which may never actually occur in the lifetime of the scheme. Current design practice for city water supplies suggests that most people would accept water rationing for a few isolated months each century rather than pay large sums of money to build reservoirs sufficient to tide them over the most extreme drought likely to occur. No one likes to be flooded, but in reality, the cost of drains designed to accommodate extreme storms and avoid damage may far outweigh the benefits. So there are choices to be made about the acceptable level of risk. Spending several million pounds on flood defences to protect a few homes may not make financial sense and it may be cheaper to move the houses. But decisions are not always as logical and simple as that and politics and perception will play a part.

Ironically, designing for droughts is more of a problem in temperate rainfall areas than in dry regions. In places where it rains, droughts may be infrequent and so the level of drought risk will be determined by assessing costs and benefits – a balance between those who want a secure water supply at all times and authorities that want to keep costs down. In contrast, designing for drought in dry regions is more straightforward. Droughts are a normal part of everyday life and so people will expect supply systems that cope with the extreme condition of no rainfall every year. Rainfall is just a bonus.

Other infrastructure, such flood alleviation schemes, bridges, dams and spillways, whose failure could result in serious loss of life, will be designed to cope with extreme flood flows. But it is not practically possible to design schemes and structures for the rarest and most extreme rainfall events. So the risk of extreme flooding cannot be eliminated. It is rare but ever present.

BOX 9.2   THE DESIGN YEAR DILEMMA FOR IRRIGATION FARMERS IN THE UNITED KINGDOM

In the UK, farmers who irrigate high-value, water-sensitive crops such as potatoes, fruits and vegetables, build reservoirs to store winter rainfall, which is usually plentiful, to irrigate crops in the summer when rainfall is less reliable. But a severe and unexpected winter drought in 2011 meant that many farmers were unable to fill their reservoirs, and some could not risk growing irrigated summer crops in 2012. Many lost money as a result and some are now thinking to reduce drought risk by increasing their storage capacity to cover two irrigation seasons rather than just one season. This is a very risky business. If it rains all through the growing season and irrigation is not required, investment in storage is wasted. But if it is dry then the ability to irrigate crops can repay farmers handsomely.

Similarly, deciding on how much irrigation equipment to buy brings more risk. Farmers expect some rainfall during the growing season and so they only invest in enough equipment to irrigate adequately 15 years out of 20 years. So the likelihood is that 5 years in 20 years they may struggle to adequately irrigate their crops even if they have enough water storage. Buying more equipment will reduce the risk but increase farm costs and reduce profits, particularly if there is a run of wetter growing seasons. So farmers in the UK, and in other supplementary rainfall areas, tend to under irrigate simply because they do not have the capacity to meet the full crop water requirements.

9.2.4 'Easy' and 'difficult' hydrology

Assessing flood and drought risks is generally easier in the richer countries across Europe than in the poorer developing countries across Africa and South Asia. In Europe, we talk about *easy hydrology*. Annual rainfall is more reliable, it is mostly within 15%–20% of the long-term average, and intensities are modest. In such circumstances, infrastructure can be designed to cope with the most likely flood and drought events. Also European countries are wealthy enough to 'engineer' their way out of most water problems.

In the developing countries, where the climate is semi-arid, we talk about *difficult hydrology*. Rainfall is seasonal, unreliable and intensive like the monsoon rains, and can vary by more than 40%–50% of the long-term average. River levels can rise rapidly washing away bridges and overtopping dams. Designing hydraulic infrastructure in such circumstances is not only fraught with engineering difficulties but structures can also simply cost too much for countries to afford them, particularly if there is a high risk they will be washed away. Kariba Dam in the Zambesi River Basin between Zimbabwe and Zambia had a major setback during early construction when extreme floods washed away the foundations which had to be reconstructed at considerable expense.

A much less catastrophic event, though no less dramatic, was an intense tropical storm across Kruger Safari Park in South Africa (witnessed by the author). The Crocodile River, which runs along the park's southern boundary, rose by some 9 m overnight. It lived up to its name as the rising water brought crocodiles from a safe place down in the river valley, up to within a few metres of the local tourist hotel!

9.2.5 Hydrographs

When a particular stretch of river is prone to flooding, engineers, who are looking to alleviate the problem, will want to gather information on how river discharges and water levels change over time. A river in flood is like a long wave which gradually moves downstream (Figure 9.5). When the wave reaches the area at risk, the water level begins to rise as the discharge increases. It rises to a peak and then subsides as the flood wave passes. Plotting a graph of discharge against water level at the chosen site produces what we call a *hydrograph*. Figure 9.5 shows hydrographs at three locations along the river as a flood wave passes. The area under each curve represents the volume of floodwater passing that point. Notice how the shape of the hydrograph changes as the flood moves downstream. The flood attenuates, which means the maximum discharge is lower downstream but the peak lasts longer. The hydrographs provide important information for assessing the level of protection needed

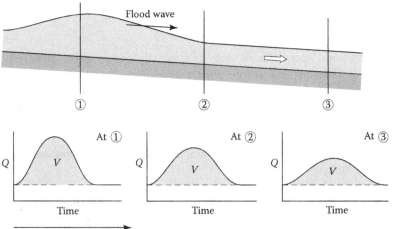

The volume of water passing each point, 1, 2 and 3 is the same. But the peak discharge is reduced and the time of flooding increases to maintain continuity.

Figure 9.5 Flood hydrographs at three locations along a river.

at different locations. Note that the volume of water passing each measuring point is the same – remember the continuity equation.

9.2.6 Climate change

Climate change has added additional uncertainty for hydrologists and engineers – is it likely to increase or decrease rainfall and/or increase transpiration if temperatures rise?

Scientists now suggest that the traditional engineering approach of looking at past rainfall records and river flows may no longer be helpful in predicting future events. But at least if the record is long enough, and in the absence of reliable information about the future, it can provide some good indication of what may happen. Most countries have experienced severe floods and droughts for many centuries. Such events are not new to the twenty-first century but the data may not be so reliable. Others suggest that the uncertainties and impacts of socio-economic change – increasing population, the desire for a more water-rich diet and lifestyle and changes in energy and industrial water demands, far outweighed the predicted uncertainties of climate change. Planning only with climate change in mind risks building infrastructure that may not be needed, particularly if the climate does not change as predicted. But planning for a range of socio-economic futures which take account of climate change may be more preferable. This is known as the 'no-regrets' option – it is something we would do even if the climate did not change, which makes a lot of sense.

9.3 DO WE HAVE ENOUGH WATER?

The big question for water resource planners is – do we have enough water? This is a question we need to ask at different scales – do we have enough globally, nationally and locally? Globally, we may decide there is enough but if we live in a dry area, such as sub-Saharan Africa it is not much comfort to know that that there is more than enough rainfall in northern Europe. In the United Kingdom, there are concerns about water availability in the drier region of East Anglia, while there is excess water in Wales and the north of England and Scotland. Unless we can physically transfer the excess, this information is of little help to those in the drier east. But let's look at the global picture and ask – do we have enough water? (See Further Reading.)

9.3.1 How much water do we use?

We have most control over water in our rivers, lakes and underground aquifers. If you can imagine a cubic kilometre of water, globally (in 2008) we withdrew 1,740 km^3 (cubic kilometres) of water. Withdrawals for

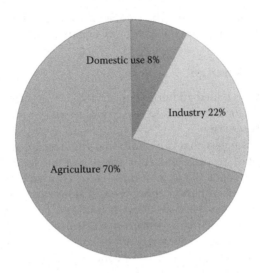

Figure 9.6 Global water use.

agriculture accounted for 1,270 km³ (70%) of this (Figure 9.6). The remaining 30% was for industry (22%) and for domestic use (8%). Agriculture, mostly irrigation, consumes the lion's share of available water resources. Domestic and industrial water use is modest in comparison.

Looking globally provides one picture of water use, but circumstances vary considerably across regions and countries. In semi-arid areas, irrigated cropping consumes as much as 95% of available water resources. Even in Europe, irrigation accounts for 44% of all withdrawals. In the UK, irrigated agriculture consumes only 1%–2% of withdrawals. But this is a national average which hides the fact that on a warm summer day in the drier East Anglia region, over 50% of water withdrawals will be for irrigating crops.

By far, the world's biggest water consumers are India and China who together consume 25% of the world's available water resources. As these two countries are growing in terms of population and economies, they are likely to consume much more in the future – a worry for us all. The United States is some way behind at 9%.

9.3.2 How much do we need?

Predicting future water demand is fraught with difficulties. It will depend on population growth, on changing life styles and eating habits, economic development, and changing environmental needs. The threat of climate change only adds to future uncertainties.

So far, the global community has not been good at forecasting the future. In 1946 IBM, an international computing company, forecast that the world would need maybe five computers; admittedly in those early days, computers were as large as houses. Forecasts made in the early 2000s have already proved inaccurate because no one seriously predicted the rise in energy prices nor the world recession and the impact these factors had on food prices. So there are questions about just how much can we rely on predictions of future water needs.

What we have some idea about is that world population is predicted to grow rapidly from 6.7 billion in 2008 to 8.3 billion in 2030 and 9.1 billion by 2050 based on UN estimates. But growth will not be evenly spread. The figures hide the fact that most growth will be in the poorer developing countries in sub-Saharan Africa and South Asia. By 2050, about 7.5 billion people will be living in low- and middle-income countries which already have limited water resources and are finding it difficult to support even today's populations.

In the higher income countries, populations are expected to decline. But the demand for water may increase as people use more water in their homes, grow more water demanding food and seek better life styles and leisure activities in an improving environment.

9.3.3 Do we have enough?

If we believe future population forecasts and continue with our current water management practices, expert opinion suggests that global water demand will double by 2050 and this will exceed available resources by more than 40%. So clearly the answer is 'no'. However, expert opinion rather suggests a cautious 'yes'. But only if we significantly change the way in which we currently use and manage water resources. So what are the options for change?

9.3.4 Strategies for matching supply and demand

We could increase the supply of water by investing in more water infrastructure, but many of the easy options have already been exploited and new sources are likely to be costly to develop, particularly in the developing countries where most of the new water will be needed.

The main strategy for bringing supply and demand into balance is to introduce *water demand management* (WDM). This is a major shift from our conventional approach to water management which looks mainly at the supply side of the equation – if water is in short supply, then increase the supply. WDM involves introducing mechanisms to reduce demand, such as water conservation practices and reducing water wastage. It is a 'carrot-and-stick' approach. It means making people more aware of the importance of water in their daily lives, discouraging waste and how misusing water can impact other users – water management thus becomes a responsibility we all share.

The 'stick' approach is to introduce water metering, increase water charges and introduce water markets so that limited water resources are used most productively. In doing this, water becomes an 'economic commodity' like other natural resources but this is where the story gets complicated. Not everyone can afford to pay for water, more so in the developing countries but even across Europe and the United States. Access to water is now recognised as a human right, but in accepting this, someone has to pay. So getting the balance right between using economic and legal tools to limit water use and protecting individual's rights to water is not an easy one.

The amount of water abstracted for domestic use is relatively small in comparison to amount consumed by agriculture. So perhaps the greatest opportunities for changing water demand will be in agriculture which currently withdraws 70% of all available water resources. Genetics may provide some help by modifying crops so they consume less water and increase crop production per cubic metre of water. Some people suggest that we change our diet from meat protein, which consumes large quantities of water, to starch-based foods, which are less water demanding. In other words, we all become vegetarians! – not a popular move perhaps (see Section 9.4).

Another food area which has the potential to reduce water demand is food waste. Remarkably, almost half the food we produce is wasted, which means that almost half of the water used to grow it is also wasted. In the developed world, food is wasted after marketing. The supermarkets reject produce because it does not meet their quality standards and we, the customers, throw food away because we bought too much or it did not look so good. In the developing world, farmers lose almost half their crops before they even get to the market because of poor long-term storage and transport facilities. So reducing food waste could have a substantial impact on water resources availability.

If reducing water for agriculture is so obvious, why has it not been done already? Unfortunately, water for food production is like the proverbial 'elephant in the room' of water resources planning, it is largely ignored. One problem is that few people realise just how much water it takes to grow our food. Another is that international attention has always been about the lack of water supply and sanitation. This is understandable because of the devastating effect that poor water supplies and sanitation have on public health and mortality and the fact that we can only survive for a few days without drinking water. But even a small reduction in the percentage of water consumed in producing our food could help to resolve many of our future water resources problems. So it is time to turn the international spotlight on water for food.

9.4 WATER FOOTPRINTS

Water footprints can help us to understand just how much water we use. It is a relatively new term and only entered the water vocabulary in 2002. It follows other footprints such as the ecological footprint which tells us how

Figure 9.7 Water for a cup of tea. 0.2 L of hot water, 5 L to grow the tea, 18 L to grow the grass and 9.2 L to grow the sugar cane. Total 32.4 L.

many earth-like planets we would need if the global population all adopted a European lifestyle (3.75 in 2015!).

The water footprint tells us how much of the world's freshwater it takes to produce goods and services. Making a cup of tea for example consumes a lot more than the 0.2 L in the cup (Figure 9.7). It takes 32 L when we account for the water to produce the tea, milk and sugar. Coffee drinkers are consuming almost five times as much (140 L), so take your pick.

Some food products have much larger water footprints. It takes 15,000 L to produce 1 kg of beef and 3,400 L to produce 1 kg of rice. Potatoes are much less thirsty at 287 L/kg.

A typical European diet requires about 3,500 L/person/day and a meat-rich diet requires as much as 5,000 L/person/day. These are quite startling figures. In contrast, a typical (poor) diet in a developing country would consume only 1,000 L/person/day. Clearly, there is a remarkable amount of water embedded in the food we eat, and it dwarfs the 150 L/person/day that we use in our homes in the UK.

9.4.1 Green and blue water

Some people use the water consumed in producing meat as a good argument for changing our diets and becoming vegetarians. But beware, the figures can be misleading. They are useful to attract attention to the concerns about water resources but they are over-simplistic indicators and do not differentiate between what we call *green water* and *blue water*.

Green water describes water from natural rainfall which is stored in the soil and used by vegetation.

Blue water refers to freshwater which finds its way into rivers, lakes and groundwater.

Why do we need this distinction? Blue water is of most interest to engineers. It is valuable and flexible and can be exploited for many different purposes, such as water supply, irrigation, energy generation and maintaining our aquatic ecosystems. Often the same flow can be used several times over. For example, water used to generate hydropower can be used again to supply households, and the return flows used downstream to irrigate crops or maintain valued wetlands. Green water is still an important resource but it is much less flexible and can only be used once where it falls for growing agricultural crops, forests and natural rangeland grazing. It is regularly replenished with rainfall and provides 60% of the world's food production.

Going back to the water footprint examples, most beef is produced using green water because cattle graze on rainfed pastures. As there is little or no alternative use for this green water, should we worry about beef production? Perhaps not, but it would be a different story if the pastures were irrigated with blue water, which has many alternative and higher value uses and may well be in short supply. Rice also relies on 80% green water and only 20% blue water.

So are water footprints useful? A water footprint is like describing a shoe only by its size. Useful information but not nearly enough. Is it leather or canvas? Is it a sports shoe or for everyday wear? Much more information is needed about the use of blue and green water and its impact on local water resources if the footprint is have meaning. Imagine trying to capture all this information on a food label in a way that the average shopper can understand if the product has used local water resources wisely.

9.4.2 Virtual water

Another widely use term is *virtual water*. It is the same as the water footprint but virtual water is used to describe the way in which water can be transported around the world. An arid country, for example, with limited blue and green water resources may decide that it can no longer grow enough food for its population without over-exploiting groundwater resources. It could import water but this would make little sense for irrigated cropping, it is heavy, bulky and expensive to transport. Much better to import the food from a country which has plenty of green water for growing crops – 1 ton of grain, for example, is equivalent to 1,000 tons of water.

Unfortunately, this simple logic does not always work in practice. Most countries like to be self-sufficient in food and not depend on others for their food security. So they tend to over-exploit their water resources, which may be okay in the short-term but not so good in the long-term. Some water-short countries even export high-value crops, grown with precious blue water resources. Such are the commercial pressures on farming to earn foreign exchange and create rural employment. But the short-term gain will have long-lasting consequences for sustaining water resources.

The UK, although a relatively wet country, imports over 50% of its food and is in effect importing about 60 km^3 of virtual water annually – some 60% of the UK's annual water footprint. Some imports come from countries that can ill afford to export food because they already face acute water shortages – tomatoes from Morocco, strawberries from Spain and potatoes from Egypt. The UK is a rich country and can afford to do this, but it does raise questions about the efficacy of importing virtual water from water-short countries. In effect, the UK is exporting its environmental problems to countries which have less rigorous environmental regulation.

9.5 PLANNING WATER RESOURCES

9.5.1 Traditional approach

Planning and managing water resources, even as recent as the 1950s, were relatively straightforward activities. In most developed countries, water was plentiful and the rules for sharing it were simple. Governments took a 'top-down' approach to solve water shortages. They increased the supply by building more dams and reservoirs and drilling wells to exploit groundwater. The word 'environment' did not exist in the vocabulary. In the UK, licences for water abstraction were issued more on the basis of 'how much do you want' rather than on what was available, with rights to abstract continuing in perpetuity. Large dams and reservoirs, built at that time, and from which we now derive great benefit, would not now be built without many more years of study, planning and enquiry.

In southern France, prosperity depends largely on the highly successful Société du Canal de Provence, set up in the 1950s, to provide hydropower and water for 2 million people, and water to irrigate 6,000 farms and run 500 industries. A French engineer, involved in the construction of this immense enterprise said that if this project was suggested today, it would be unlikely to get planning approval because of its potential environmental impact and the many objections from the various environmental interest groups. This maybe a 'tongue-in-cheek' comment but it highlights the ever-increasing complexity of planning and managing water resources today.

Governments usually have the responsibility for the overall management and protection of a nation's water resources. But there are many public and private organisations that use water. Typically, a ministry of public works has responsibility for managing domestic and industrial water supplies and sanitation; a ministry of water and energy manages water for generating hydropower and cooling power stations, and a ministry of agriculture and environment is responsible for managing water for agriculture and irrigation, and the aquatic environment. Historically, ministries have focused on their particular tasks to the exclusion of other water interests even though they are all exploiting the same resource. But as long as water was plentiful, this fragmented 'silo' approach was not a problem.

9.5.2 Integrated water resources management

Problems only come when there are water shortages and serious conflicts arise among users. This is often the catalyst for change as there is nothing like a good crisis to bring people out of their silos and get them talking and working together to find solutions. This was the beginning of the movement towards a more integrated approach to managing water. The idea being that if the water sector and water users worked together to plan water for people, industry, food, energy and the environment, we can amicably negotiate the trade-offs needed to keep supply and demand in balance. It is of course in everyone's long-term interest to do this. This is a simple and compelling idea and hard to disagree with. But many have found it difficult to put into practice.

At first, integration meant bringing together water resources with engineering and economically driven solutions for efficiency gains and more coordinated decision-making. An early example was the Tennessee Valley Authority (TVA) in the United States which, in 1933, brought together under one roof the different facets of water use for economic development, such as navigation, flood control and power generation.

In 1992, as concerns worldwide began to grow about water shortages, much more comprehensive plans for integration were proposed that went beyond economic development and began to embrace the wider social and environmental aspects of water management. These became what are now known as the UN 'Dublin Principles'. These established the concept of Integrated Water Resources Management (IWRM). The Global Water Partnership, an organisation set up in 1995 to promote IWRM, defined it as: *a process which promotes the coordinated development and management of water, land, and related resources in order to maximise economic and social welfare in an equitable manner without compromising the sustainability of vital ecosystems.* IWRM is about achieving strong economic growth, equitable and fair allocation of water, and protecting the environment.

Technically, IWRM recognises two important points. First, that planning and management should take place at the water basin level. This is both logical and sensible from a hydrological point of view (see Section 9.2). Second, land management is important to successful water management. Water runs across the land to get to the rivers. So what happens on the land influences the quantity of water in the river and also its quality. The way the land is used will determine how much water runs off. Planting forests may be one way of slowing down flood flows but equally forests need water to grow and this will impact summers flows in the basin. Where the land is cropped, agricultural chemicals can find their way into the rivers and affect water quality for other users. In turn, water quality can affect how we use the land to produce food crops.

Since 1992, water management has gradually shifted from the 'top-down-supply-more-water' approach to one that focuses on getting the best deal out of the available water resources we have. This means focusing on ways of reducing water demand (see Section 9.3), giving more attention to the natural environment which sustains our current water resources and enabling people (stakeholders) to have a much greater say in how we do things. In other words, water management has got a lot more complicated.

IWRM is not without its critics and many have found it difficult to put it into practice. It was criticised as being ill-defined and left people wondering how to do it in practice. But like many complex issues, people are finding ways of introducing it and indeed the United Nations has enshrined IWRM within the 2015 development agenda in its SDG 6, the Water Goal, which says that *by 2030 we must implement integrated water resources management at all levels, including transboundary cooperation*.

IWRM is described as a process and not a 'blueprint' solution for water management. It is not like building a dam, and once people realise this, they can stop looking for that one solution. Water resources are different from place to place and so too are the development priorities and social and economic issues. So there is no one single answer to the question of 'how to integrate'; it depends on local circumstances and each country must find its own way of doing things that are best suited to them.

One stumbling block concerns the prerequisites for IWRM to succeed. These include a strong enabling environment in the country, sound investments in infrastructure, clear and robust institutions and an effective range of technical and managerial tools to do the job. Most of these are in place in the developed world but many of these elements are sadly lacking in the developing countries. So there is a lot of preliminary work to be done before IWRM has a chance of succeeding.

9.5.3 Water security

Even though IWRM is now centre stage within the United Nations, it is considered by some to put too much attention on the process of managing water rather what we are trying to achieve, which is more secure water for everyone – for people, for food and energy production and for the environment. For this reason, the idea of *water security* has gained more acceptance than IWRM and encourages not only the water managers but also the water users to become involved in water management. Thus, IWRM becomes the means of achieving water security.

The aims of both are identical but water security is a more intuitive term and one that most people are identifying with even if they view the issues and priorities differently. Sometimes the water community gets hung up on terminology rather than finding solutions to problems. Social scientists suggest that a common language is important for resolving complex problems and as such 'water security' is a language that is bringing people together to talk and to plan for the future.

Engineers and scientists are now investigating ways of measuring water security. Remember the words of the physicist, Lord Kelvin (1824–1907) – it is essential to put numbers on things if we really want to understand them. But how do you put a number on water security. Well some people are trying to do just that. So lots of exciting times ahead for those involved in water management!

9.6 WATER MYTHS

This final section deals with water myths (Figure 9.8). The crisis predicted over future water resources has meant that myriad national and international organisations are all trying to get involved in water. This is a good thing, but (cynically) they all also see lots of opportunities for funding their projects. Governments, international agencies, research organisations, universities, commercial businesses, civil society and non-governmental organisations, all want to have a slice of the action and have their say.

But many are new to the world of water and are not so familiar with water as a natural resource – where it comes from, where it goes after use, its physical and chemical properties and how water behaves when it is still and when it starts to move. The solid world around us is more familiar and so we tend to rely on this experience to help us understand the world of water. But how water behaves in practice is not always intuitive. Remember the big difference between moving crowds of people along corridors and through narrow doorways, and moving water through narrow pipes and channels (see Section 3.7.1). Remember also how it was essential to narrow the opening under a bridge across a river to lower water levels and reduce flooding, when all our instincts told us to make it bigger (see Section 5.7.1).

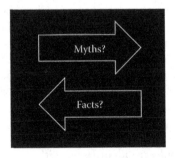

Figure 9.8 Myths or facts?

And remember how an understanding of hydraulics comforts us with the knowledge that we cannot actually drown in quicksand – unless we go in headfirst of course (see Section 2.12.4).

To work with water is to understand that water can behave differently to our solid world and if we do not want to make some big mistakes about how we manage water better in the future, we need to be aware of a number of important water myths. Here are some of them.

9.6.1 Myth I: Water resources are finite

We are told regularly by media and politicians that water is a finite resource as there is only so much of it on the earth. This is true, but it would be quite wrong to think about water as finite in the sense that fossil fuels are finite. We are not going to run out of water like we might one day run out of oil.

Once we consume fossil fuels, they are gone and we cannot use them again. Water is different – it is renewed annually. It is continually moving, emptying and replenishing. Water availability is limited only by the amount of rainfall, which varies over time, and our ability to capture, control, store it and use it wisely. In some circumstances, we can use river water many times over before it finds its way back into the sea. Each litre of water in the River Thames is reputed to go through about five different people before it finally reaches the sea. So 1 L of water becomes 5 L in accounting for water availability.

If we want more water, we can build more reservoirs, recharge aquifers and increase soilwater storage. This is a cost issue and not a water issue. If we are prepared to spend more, we can have more water. As one colleague put it *the UK is an island surrounded by water. All we have to do is to clean it up*. This option may be financially and environmentally unpalatable to our society but it is there if we are prepared to pay for it.

Thinking about water as a finite resource can lead to some poor and ineffective decision-making. A good example of this is the current spending on increasing water use efficiency in both irrigated agriculture and municipal water supplies (see Myth III). It is much better to think in terms of water as a limited resource. This means we understand the water balance and the hydrological cycle and so we are likely to make more sensible decisions about how we plan and manage water resources.

9.6.2 Myth II: Water use and consumption are the same

Water use and consumptive use have very different meanings yet they are usually lumped together to describe water demand. This is really like counting apples and oranges as the same fruit. We *use* water for domestic and industrial purposes, and once we are done with it, we put it back into the drains so that it can be cleaned up and used again. The recycled water becomes a resource for others to use. Water used for irrigation and for environmental purposes is *consumed*. The water is transpired through vegetation into the atmosphere and is lost to the immediate hydrological cycle. There is no flow back into the soil once the crops get hold of it. Water for power stations is a mixture of use and consumption as some is evaporated during cooling processes. So separating use and consumption for the purposes of planning and management makes a lot of sense.

Indeed, some engineers suggest that water allocation and management should be based on consumption rather than abstraction.

One poster in a public bar in the UK told customers that they were only using the beer and not consuming it. The landlord expected all customers to return the beer before leaving the bar!

9.6.3 Myth III: Increasing water-use efficiency saves water

Most people are familiar with the term *efficiency* and what it means. Usually, it is about using less of some resource and the implication is that this is a good thing to do. We insulate our homes to reduce the amount of energy we consume and drive fuel-efficient car that consume less petrol. But this rather simple idea does not transfer easily to water and using less water is not always the same as saving water. Here are some examples to illustrate this.

9.6.3.1 Individual efficiency versus basin efficiency

Many irrigation schemes in developing countries have a reputation for poor water use efficiency. This is measured by dividing the amount of water needed to grow the crops by the amount diverted into the scheme from the source. Headline figures showed that most schemes are only 50% efficient at best, which implies that over half the water turned into the scheme is wasted or lost. This was one of the reasons why international development agencies lost interest in investing in irrigation schemes in the 1980s. But is this a fair measure of what is happening?

Consider this example: A group of farmers take water from a river for irrigation. Each is operating their irrigation at only 50% efficiency – so half the water that each of them abstracts is not consumed by their crops and goes to waste (Figure 9.9). Now assume that the first farmer takes 1,000 units of water. He consumes 500 units and wastes 500, which runs off into the drains, flows back into the river and becomes a source of supply for the next farmer downstream. He abstracts 500, uses 250 and because he is also working at 50% efficiency, 250 units flow back into the river. The third farmer abstracts 250 and wastes 125 and so on. At this point, of the 1,000 units originally abstracted from the river, a total of 875 units

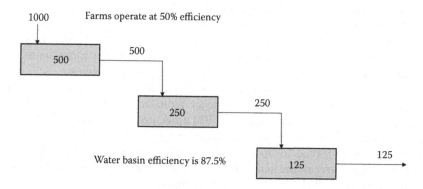

Figure 9.9 Sharing water along a river.

(500 + 250 + 125 units) are usefully consumed by the crops, albeit on different farms. So although each farmer was operating at only 50% efficiency, the overall efficiency of water use by the group of farmers is actually much higher at 87.5%. This is a much more acceptable level of efficiency.

9.6.3.2 What can we conclude from this?

- When water is 'lost' from an individual farm it is not always lost to the basin. It is still there for someone else to use downstream. This is very different to what happens to energy. When energy is lost to one home because of poor insulation, it cannot be used by another home nearby.
- If there are opportunities for water users downstream, spending money on improving the efficiency of individual upstream farms may not be worthwhile. As one observer pointed out 'increasing efficiency is always costly but not always beneficial'.
- If there are no beneficiaries downstream, then again is there much sense in farmers spending money to improve efficiency if the basin is open?
- It can be argued that increasing the efficiency of individual farms will reduce the farm water bill and the energy costs in pumping water which goes to waste. This is true and maybe worthwhile. But the incentive for increasing efficiency is not water-saving but money-saving, which is what farmers are most interested in.
- Scale matters when assessing water use efficiency. Individual water users are likely to invest in efficiency measures only if it saves them money. Water basin managers should not be interested in individual farms but focus on farmer groups along a river or the basin as a whole. If the group or basin efficiency is high, because of reuse opportunities, they should be happy and not worry too much about what individuals are doing.

Clearly, this is a simplified model and if circumstances change, then the decision-making along the river will change. If the source was groundwater, farmers are not connected downstream in the same way as before. In such circumstance, addressing individual farm efficiency makes more sense, as farmers may be over-abstracting precious groundwater sources.

One final point, if for some reason the first farmer increases his efficiency to say 100% (for simplicity). Experience from across the world suggests that the farmer would still abstract 1,000 units, grow more crops and make more money. He would be unlikely to offer 500 back as a gift to the environment unless forced to do so. So investing in farm irrigation efficiency may not save water in the sense of making more available for someone else to use, it is more likely to increase farm profits.

9.6.3.3 Improving 'efficiency' for whom?

There are many examples around the world of questionable investment in water use efficiency. Here is just one. Efficiency is often the driving force for change but it is not always clear who benefits. The 'All American Canal' runs along the border between the United States and Mexico and provides irrigation water for the Imperial Valley, a rich agricultural area. But seepage from the canal meant that almost half the water was 'lost' in transit. The solution was to line the canal with concrete. This increased the supply and lots of US farmers were very happy. But south of the border in Mexico, many thousands of smallholder farmers were very unhappy because their irrigation wells began to dry up and many lost their livelihoods as a result. For years, they had come to rely on pumping local shallow groundwater to grow crops. The groundwater was, of course the result of seepage from the canal. The US farmers increased their crop production and profits but at the cost of the livelihoods of many thousands of Mexican small farmers. So how do you assess water use efficiency of the canal in this situation? It all depends whether you are a US or a Mexican farmer.

9.6.3.4 Domestic water systems

This same principle can be applied to domestic water supply systems. In the UK, the average person uses about 150 L of water each day for domestic purposes. Water companies are keen to reduce this amount to say 130 L by introducing efficiency measures in homes such as installing water meters, reduced flush toilets and showers to replace baths. The idea is that such steps will make more water available for others to use as the population grows. In principle, this works fine at this local level; people install 'efficiency' devices and reduce their water bills, which is probably the main incentive for them making the investment. But it could have impacts at the larger scale. If there are downstream users who rely on the wastewater outflow from upstream homes for their supply, which is quite common

---

BOX 9.3   ONE FAMILY REPORTS SIGNIFICANT WATER SAVINGS

This story was reported in the UK civil engineering press (rather tongue in cheek I think):

Following an efficiency drive by a water company, one water engineer reported that his family was able to reduce their daily water use from 150 to less than 100 L/person/day – a remarkable achievement. How did they do it? Well it turned out they visited their local swimming centre to shower in the morning and to use the toilets; they washed their clothes at the launderette and they ate their meals in local restaurants.

Clearly, this family used just as much, if not more water, by doing this. They 'saved' water and reduced their household water bill, but may well have spent more eating out. Importantly, there was no overall water saving for the water company. There was no 'extra' water made available for other customers to use. This is what the US water engineers call *dry water savings* as opposed to *wet water savings* which really do make more water available for others.

Simply basing the measure of efficiency on water going through the household water meter is clearly not enough! And it again demonstrates the interconnectedness of water and the dangers of using over-simplistic measures to show what is happening.

(see Section 8.3), there will be less water available for them to use. So the actions of a local group of individuals to reduce their water bills will mean that users downstream, who may have a different water supplier with different rules, will now be short of water (see Box 9.3).

The implication of this is that upstream water companies cannot operate completely independently of other companies downstream. If the company at the head of the system changes the allocation rules, they will need to discuss and agree these with those downstream. Water connects us all in ways that other resources, like energy do not.

9.6.4 Myth IV: Wastewater is the 'new' water resource

Treated wastewater (see Section 8.4) is often 'discovered' by people and politicians who believe they have found a 'new' water resource which no one has yet thought about. But further investigation often shows that this resource is already being well used. In developed countries, it is mixed with freshwater, following treatment and diluted prior to reuse for domestic purposes or used directly for irrigating crops. In many developing countries, there is often not the luxury of treatment and dilution but it is still used again as some people have no other supply options.

In some countries, effluent is discharged both treated and untreated into sea. This truly is a waste. In such cases, this water can be turned around and become a 'new' resource. But unfortunately, the rest is not just lying around waiting for someone to discover it. It is already being well used.

9.6.5 Myth V: Societies should only use water they can replace

The current mantra of sustainability suggests that we should only use and consume water that we can replace in order not to degrade our water resources and aquatic ecosystems. This is a fine sentiment in Western Europe, but in arid regions, there are not many sustainable options when countries want to grow their economies. Small farming communities across North Africa and the Near East sustain their livelihoods using flash floods and water harvesting to grow crops, and exploit (often over-exploit) local groundwater for domestic use and for irrigation. Where conditions are truly arid, people rely on livestock herding with goats and camels that can go for long periods without much water. But all this is low-level subsistence farming. It may be sustainable, but it is not about economic development and improving livelihoods and life quality. One option across much of North Africa and the Arabian Peninsula is to exploit fossil water (groundwater laid down thousands of years ago) for economic and social development. Rather like we consume fossil fuel resources, this water is not replaceable. Some condemn such practices, but in doing so, they condemn people to a subsistence lifestyle. The argument from one Libyan water engineer, as Libya sits on top of one of the world's largest aquifers, is *'do not condemn us to a life scraping a living with goats in the desert. There is over a 1,000 years supply of water down there and if we cannot come up with an alternative source in the next millennium then we do not deserve to continue living in this place'*. So perhaps we should not be too quick to condemn and impose western thinking on others living in very different circumstances.

Chapter 10

'Bathtub' hydraulics

This final chapter is all about having a bath. It is a really good practical way to learn about hydraulics and can provide you with inspirational moments. Here are just some of the things to look out for.

Just filling the bath can be an experience in itself. The hot and cold water taps are running but then you notice that the two are not mixing well. One side of the bath is hot and the other cold. There is some mixing at the interface due to turbulence but this is all. The only way to get an even temperature is to stir the water vigorously with your hand. The reason for this mixing problem is that water density varies slightly with temperature and this density difference is enough to inhibit mixing. This is a major problem at most power generating stations that use vast quantities of water from rivers and the sea to cool their systems. High towers help to cool water before it is returned to the river or sea but any slight difference in temperature will stop it mixing fully with the receiving flow. Swimming downstream of a power station can be a pleasant, warm experience. The challenge for the engineers is to find ways of mixing the water thoroughly so that the receiving water returns to its original temperature as quickly as possible and does not affect local aquatic plant and fish life. They also need to stop hot water short-circuiting the system and finding its way into the intake and back into the power station as can happen with coastal stations.

As water flows from the taps across the bottom of the bath the flow is usually supercritical – fast and shallow. But this state soon changes as a hydraulic jump forms at the far end and then quickly makes its way back towards the tap end as a traveling surge wave. For a while a stable circular hydraulic jump can be seen just where the water plunges into the bath but eventually this is drowned out as the level in the bath rises. The incoming flow is still supercritical and this now shoots under the slow flow. The energy is not dispersed in the hydraulic jump; it is gradually absorbed as friction along the base of the bath slows the water down. When this occurs in natural channels it can cause severe erosion of the bed and sides (Section 6.2).

Once your bath is full to the right level and is at the right temperature, there is now that 'Eureka moment' which Archimedes experienced when he stepped into the public baths in Syracuse. Archimedes first discovered the significance of this some 2000 years ago when he realised that the water displaced when you get into the bath has the same volume as your body. He also noticed that if you float on the water instead of sitting on the bottom then the amount of water you displace is equal to your weight. From this he was able to solve the problem that the King of Syracuse had over what materials had been used to make his crown – was it gold or was it really lead dressed up to look like gold (Section 2.12)? Recent research into how people have bright ideas suggests that the best thoughts do not always come to you when you are working hard and focusing on the problem. Apparently, the best ideas pop into your head when you least expect them; when walking in the park with your dog, or taking a pleasant relaxing bath. Is this how the idea came to Archimedes?

This was the beginning of our understanding of hydrostatics (water which is not moving) and led to formulae that we still use today to design water tanks, dams and submarines. It is an almost perfect theory. It was unfortunate that those early scientists tried their hand at hydrodynamics – water which is moving – but they got this wrong and sent science off in the wrong direction for almost 2000 years. We had to wait until Sir Isaac Newton came along to put things right.

Sitting still in a bath, soaking up the warmth and hoping for inspiration is a good experience – but for children this is almost impossible. Sliding up and down quickly is much more fun as you can make waves and even make the water flow over the sides of the bath (Figure 10.1). As you start to move up and down the bath you transfer your body momentum to the water by surface and form drag (Section 3.10) – the larger you are the more form drag you can create and the more water you can move. When you

Figure 10.1 Waves in the bathtub.

stop sliding about, the waves seem to continue for a while. This is because the only force available to absorb the wave energy is friction and as there is very little of this in a bath it takes some time to suppress the waves. Water can also slosh about in harbours in much the same way as in the bathtub and it can cause lots of problems for loading and unloading ships. The wave energy comes from the sea, it enters the harbour, and it is difficult to get rid of it because, like the bathtub, the walls of the harbour reflect the energy rather than absorb it. The movement of the water can move ships back and forth on their moorings which can be a major problem if the ship happens to be a super tanker and you are trying to keep it still while loading it with oil. This is why harbour entrances are narrow and specially angled to stop wave energy from entering. You may have noticed that the sea is much calmer inside a harbour than outside. Some harbours though have been known to behave quite the opposite. The waves get bigger because of resonance created by the wave patterns and the size and shape of the harbour.

As you relax in the bath you decide to have a drink. A glass of whisky (or cordial) will do the trick but you may want water with it. So do you put the water in the whisky or the whisky in the water? If you put a spoonful of whisky in a glass of water and stir it and then put back one spoonful of the mixture back into the whisky; is there more whisky in the water or more water in the whisky? See if you can work that one out – the answer is in the box at the end of this chapter.

When you get out of the bath and take out the plug, a whirlpool or vortex is set up which seems to be hollow down the middle. This is a boundary layer effect (Section 3.9.3). The boundary layer close to the outlet slows the flow velocity which makes the water swirl and the vortex then forms. We say that the boundary layer curls up. All water intakes at reservoirs and control gates along rivers suffer from vortices like this which draw in air and reduce the water discharge. If you put some floating object over the vortex, such as your plastic duck, it stops the swirling and the discharge down the plughole increases. This is what engineers do in practice to stop vortices from forming at offtakes and also pump station suction inlets. They use floating or fixed grids rather than plastic ducks. Setting the outlet or the pump intake deep below the water surface will also suppress the vortices.

Finally, there is always the question of which way the vortex goes. Some say that the vortex goes in a clockwise direction in the northern hemisphere and anticlockwise in the southern hemisphere. There is a very practical demonstration of this by an enterprising young man who lives on the equator in Kenya and puts on demonstrations for the tourists. He has a large can filled with water which he sets up 20 m north of the equator. When he pulls a plug out of the bottom of the container the water slowly starts to swirl in a clockwise direction. He then does the same test 20 m south of the equator and the swirl starts in an anticlockwise direction. It works every time. Convinced? It is a good trick. The reality, however, is that no one is absolutely sure. Experiments to test the theory have been tried but they are very difficult to do in a laboratory. The problem is that the force which causes the vortex to start – the Coriolis force which comes from the earth's rotation – is very small in comparison to other forces around, such as minor vibrations due to traffic outside the laboratory or temperature changes in the room which can set up convection currents in the tank. All these can significantly influence which way the water will begin to swirl and override the effects of the Coriolis force. A large tank of water is needed to get an appreciable Coriolis force but arranging this under laboratory conditions is not very practicable.

If you prefer a shower to a bath, even here there is something to learn. When you switch on the shower and draw the plastic curtain around you for a bit of privacy, have you noticed how it tends to cling to your body, cold and uncomfortable. This is because the fast downward flow of water from your shower causes a slight drop in air pressure within the curtained space (remember the energy equation, energy changes from pressure energy to kinetic energy). The pressure outside the curtain is still at atmospheric pressure – slightly greater than the air pressure inside – and so the pressure difference causes the curtain to move towards the water and to cling to you.

So I recommend you bathe rather than shower. It is more pleasant and relaxing. You may use more water than if you had a shower, but remember it is not always wasted. It may cost you a bit more in water charges but most of the water you use goes back into the system for someone else to use again downstream. Remember Water Myth III on efficiency (Section 9.6.3). You may also experience one of those Eureka moments – just like Archimedes! Now just enjoy that drink; after you have solved the mixing problem (see Box 10.1).

BOX 10.1 EXAMPLE: A MIXING PROBLEM

Take one glass of water and an equal glass of whisky. One spoonful of whisky is put into the water and mixed. One spoonful of the mixture is put back into the whisky. Is there more whisky in the water or more water in the whisky?

Start by assuming that each glass holds 10 spoonfuls – so the water glass holds 10 spoonfuls of water and the whisky glass hold 10 spoonfuls of whisky. Now follow the argument below under the water and whisky headings as liquid is moved from one to the other:

Water	Whisky
10_{water}	10_{whisky}

One 'spoonful' (one part) of water is taken from the water and added to the whisky

9_{water}	$10_{whisky} + 1_{water}$

The whisky glass now holds 11 spoonfuls of the mix. Each spoonful of the mix comprises 10 parts water and one part whisky, that is, $(10_{whisky} + 1_{water})/11$.

Now take one spoonful of the mix and return this to the water:

$9_{water} + (10_{whisky} + 1_{water})/11$	$10_{whisky} + 1_{water} - (10_{whisky} + 1_{water})/11$
$9_{water} + 1/11_{water} + 10/11_{whisky}$	$10_{whisky} + 1_{water} - 10/11_{whisky} - 1/11_{water}$
$9.1/11_{water} + 10/11_{whisky}$	$9.1/11_{whisky} + 10/11_{water}$

So the result is, the amount of water in the whisky is the same as the amount of whisky in the water. So whichever way you mix your drinks it make no difference.

Further reading

Many of the books and references listed here may seem rather dated to some readers but nonetheless they are still excellent value in providing information for resolving hydraulic problems. The basic principles of hydraulics do not follow fashion; they remain the same – like Newton's laws of motion! Most of the texts are still available via Amazon.

BS3680. 1979. *Measurement of Liquid Flow in Open Channels*. British Standards Institution, London (also ISO 748 1979).

BS3680. 1981. Part 4A: Thin Plate Weirs.

BS3680. 1981. Part 4C: Flumes.

BS3680. 1990. Part 4E: Rectangular Broad Crested Weirs. These are the standards which are the basis for the design, installation and operation of flow measuring structures.

Chadwick A and Morfett J. 1998. *Hydraulics in Civil and Environmental Engineering*. 3rd edition. E & FN Spon, London. Similar in some ways to Webber, this is another excellent undergraduate text but its coverage of hydraulics in a much wider and includes sediment transport, river and canal engineering, coastal engineering and hydrology.

Chow VT. 1981 *Open Channel Hydraulics*. International Student edition. McGraw–Hill, New York. This is still the bible of open channel hydraulics. It is a handbook with lots of information on all aspects of channel flow and design -it is not an elementary text.

Douglas JF. 1996. *Solving Problems in Fluid Mechanics*, Parts I and II. Longman, Harlow, UK. This is a good problem-solving oriented text. It approaches the principles of hydraulics by showing how to solve numerical problems.

Global Water Partnership (GWP) This is an independent organisation, based in Sweden, that publishes extensively on IWRM and other water matters primarily for the developing countries. All publications available on www.gwp.org

Hamill L. 2001. *Understanding Hydraulics*. MacMillan Press Ltd, Basingstoke, UK. This is a book written by a teacher of hydraulics who clearly has a lot of experience of teaching a difficult subject. It is written in a conversational style with the student asking questions and the teacher providing the answers. It is sympathetic to the needs of young engineers but makes no allowances for those with limited mathematical skills.

Hydraulics Research Wallingford UK. 1982. *Charts for the Hydraulic Design of Channels and Pipes*. Thomas Telford Ltd, UK. An excellent and widely used book of design charts based on the Colebrook-White equation.

Kay M. 1983, *Sprinkler Irrigation: Equipment and Practice*. Excellent practical guide to all aspects of sprinkler irrigation systems. Out of print but available on Amazon.

Kay M. 1993, *Surface Irrigation: Systems and Practice*. Excellent practical guide to all aspects of surface irrigation systems. Out of print but available on Amazon.

King HW and Brater EF. 1996. *Handbook of Hydraulics for the Solution of Hydrostatic and Fluid Flow Problems*. 7th edition. McGraw-Hill, New York. Very useful for the solution of Manning's equation for channel flow. His method is used in this text.

Portland Cement Association. 1964. *Handbook of Concrete Culvert Pipe Hydraulics*. Portland Cement Association, Chicago. Very comprehensive book on culvert hydraulics and their construction. Although it is primarily supporting concrete culverts they can be made out of other materials as well.

Rouse H and Ince S. 1957. *History of Hydraulics*. Dover Publications, New York. A comprehensive, well researched and very readable history of hydraulics from the early work of Greek scholars to our modern notions of fluid behaviour. Rouse is a well-known authority on hydraulics.

United Nations Development Agenda 2015 and Sustainable Development Goals (SDGs). http://www.un.org/sustainabledevelopment/development-agenda/ All you need to know about the UN Development Agenda and the role that water will play.

USBR. 1974. *Design of Small Dams*. United States Dept of the Interior, Bureau of Reclamation, USA. Lots of useful construction details of hydraulic structures and particularly weirs and spillways.

Vallentine HR. 1967. *Water in the Service of Man*. Penguin Books Ltd, Harmondsworth, UK. Sadly this book is now out of print but it may still be available in some libraries. It is an excellent, easy to read, introduction to the fundamentals of water flow in pipes, channels and pumps as well as providing a broader appreciation of water and its uses including its history. The text is very descriptive, anecdotal and entertaining in style with lots of good explanations and very little mathematics. Vallentine is clearly an engineer who knows how to communicate his ideas in a practical and interesting way.

Water for Food Water for Life. 2007. An excellent review which addresses 'How much water do we need?' Earthscan, London, UK.

Webber NB. 1971. *Fluid Mechanics for Civil Engineers*. Chapman Hall, London. An excellent, comprehensive undergraduate civil engineering textbook covering both basic principles and practical applications.

World Economic Forum Report. 2015. https://www.weforum.org/agenda/2015/01/why-world-water-crises-are-a-top-global-risk/. Explains why water is now the No. 1 risk to world security.

Index

Printed in the United States
by Baker & Taylor Publisher Services